2nd Iberian Nuclear Astrophysics Meeting on Compact Stars 2011

Journal of Physics: Conference Series Volume 342

Salamanca, Spain
22-23 September 2011

ISBN: 978-1-62276-688-8
ISSN: 1742-6588

Printed from e-media with permission by:

Curran Associates, Inc.
57 Morehouse Lane
Red Hook, NY 12571

Some format issues inherent in the e-media version may also appear in this print version.

Printed by Curran Associates, Inc. (2013)

For permission requests, please contact the Institute of Physics
at the address below.

Institute of Physics
Dirac House, Temple Back
Bristol BS1 6BE UK

Phone: 44 1 17 929 7481
Fax: 44 1 17 920 0979

techtracking@iop.org

Additional copies of this publication are available from:

Curran Associates, Inc.
57 Morehouse Lane
Red Hook, NY 12571 USA
Phone: 845-758-0400
Fax: 845-758-2634
Email: curran@proceedings.com
Web: www.proceedings.com

TABLE OF CONTENTS

ORGANIZING COMMITTEE

Dr M Ángeles Pérez-García (Área Física Teórica-Universidad de Salamanca & IUFFYM)
Dr J A Miralles (Universidad de Alicante)
Dr J Pons (Universidad de Alicante)
Dr C Albertus (Área Física Nuclear-Universidad de Salamanca & IUFFYM)
Dr F Atrio (Área Física Teórica-Universidad de Salamanca & IUFFYM)

PREFACE

The second Iberian Nuclear Astrophysics meeting was held at the University of Salamanca, Spain on 22–23 September 2011. This volume contains most of the presentations delivered at this international workshop. This meeting was the second in the series following the previous I Encuentro Ibérico de Compstar, held at the University of Coimbra, Portugal in 2010.

The main purpose of this meeting was to strengthen the scientific collaboration between the participants of the Iberian and the rest of the southern European branches of the European Nuclear Astrophysics network, formerly, COMPSTAR. This ESF (European Science Foundation) supported network has been crucial in helping to make a broader audience for the the most interesting and relevant research lines being developed currently in Nuclear Astrophysics, especially related to the physics of neutron stars. It is indeed important to emphasize the need for a collaborative approach to the rest of the scientific communities so that we can reach possible new members in this interdisciplinary area and as outreach for the general public.

The program of the meeting was tailored to theoretical descriptions of the physics of neutron stars although some input from experimental observers and other condensed matter and optics areas of interest was also included. The main scientific topics included:

- Magnetic fields in compact stars
- Nuclear structure and in-medium effects in nuclear interaction
- Equation of state: from nuclear matter to quarks
- Importance of crust in the evolution of neutron stars
- Computational simulations of collapsing dense objects
- Observational phenomenology

In particular, leading experts from the computational simulation of core-collapse supernovae and the effect of hadron–quark phase transitions developed specialized review talks. Prospects in future observations or a more dilute classification of magnetars were also discussed. The importance of the equation of state, three-body forces, finite nuclei, phenomenological fermionic interaction models, and the microphysics inputs of different many-body approaches to some very important quantities as the symmetry energy were reviewed and discussed from either the non-relativistic to the relativistic framework. The importance of the crust with the existence of a crystallized structure and vortex-crust pinning were some of the important subjects discussed in the context of cooling and field dynamics.

Finally, some condensed matter and optics talks presented us the rich insight that Cold Atom Physics can give us on low-density interactions and the new and very intense laser Petawatt beams can test matter under strong external fields, respectively.

We would to thank the Faculty of Science and University of Salamanca for hosting the meeting. We also thank for partial financial support the European ERC Network COMPSTAR, The Physics of Neutron Stars under reference 3803 and the Spanish Ministerio de Ciencia e Innovación (MICINN) with project FIS2011-14759 and the local institutions of Instituto de Física Fundamental y Matemáticas (IUFFYM) and Universidad de Salamanca, Spain. Of course we thank those who have contributed to make this meeting a nice occasion to gather and start to develop fruitful collaborations. To them go our grateful acknowledgments.

December 2011, Salamanca,Spain
M Ángeles Pérez-García, J A Miralles, J Pons, C Albertus, F Atrio
Organizing Committee of II Iberian Nuclear Astrophysics Meeting

SPONSOR OR FUNDING ACKNOWLEDGMENTS

European ERC Network COMPSTAR, The Physics of Neutron Stars under meeting ref. 3803

comp
star

Spanish Ministerio de Ciencia e Innovación (MICINN) with project FIS2011-14759

MINISTERIO
DE CIENCIA
E INNOVACIÓN

Instituto de Física Fundamental y Matemáticas (IUFFYM)

IUFFyM

Universidad de Salamanca, Spain

VNiVERSiDAD
D SALAMANCA

MULTIDARK Consolider-Ingenio 2010, MICINN ref. CSD2009-00064

MultiDark
Multimessenger Approach
for Dark Matter Detection

PICTURE OF PARTICIPANTS

Picture of some of the participants of the II Iberian Nuclear Astrophysics Meeting.

LIST OF PARTICIPANTS

C Albertus (U. Salamanca, Spain)
I Bombaci (U. Pisa, Italy)
Rudiney Casali (U. Coimbra, Portugal)
Silvia Chiacchiera (U. Coimbra, Portugal)
Anthea Fantina (U. Libre Bruxelles, Belgium)
Marcio Ferreira (U. Coimbra, Portugal)
Miguel Gullón (U. Alicante, Spain)
Fabrizio Grill (U. Coimbra, Portugal)
Joe Hughto (Indiana University, USA/ U. Alicante, Spain)
J M Ibáñez (U. Valencia, Spain)
B Juliá Díaz (U. Barcelona/ICFO)
D Logoteta (U. Coimbra, Portugal)
V Moreno (U. Autónoma de Madrid, Spain)
M A Pérez-García (U. Salamanca, Spain)
J Pons (U. Alicante, Spain)
C Providencia (U. Coimbra, Portugal)
Nanda Rea (ICE-CSIC, IEEC, Barcelona, Spain)
Xavier Roca-Maza (INFN, Milano, Italy)

Luis Roso (CLPU/ U. Salamanca, Spain)
Klaas Vantournhout (GSI Darmstadt, Germany)
I Vidaña (U. Coimbra, Portugal)
Daniele Viganó (U. Alicante, Spain)

All papers published in this volume of *Journal of Physics: Conference Series* have been peer reviewed through processes administered by the proceedings Editors. Reviews were conducted by expert referees to the professional and scientific standards expected of a proceedings journal published by IOP Publishing.

Evolution of proto-neutron stars with hadron–quark phase transition

Ignazio Bombaci

Dipartimento di Fisica Università di Pisa and Istituto Nazionale di Fisica Nucleare, sezione di Pisa, largo B. Pontecorvo, 3, 56126, Pisa, Italy

E-mail: bombaci@df.unipi.it

Domenico Logoteta, Constança Providencia, and Isaac Vidaña

Centro de Física Computacional, Department of Physics, University of Coimbra, 3004-516 Coimbra, Portugal

Abstract. We investigate the consequences of the quark deconfinement phase transition on the evolution of proto-neutron stars. Assuming a first order phase transition, we calculate the nucleation time due to thermal and quantum nucleation mechanisms. We introduce the concepts of critical mass M_{cr} and limiting conversion temperature for proto-hadronic stars. We show that proto-hadronic stars with a mass $M < M_{cr}$ could survive the early stages of their evolution without decaying to quark stars.

1. Introduction

The core of a neutron star is one of the best candidates in the universe where a phase of strong interacting matter with deconfined quarks could be found. Thus these compact stars can be viewed as natural laboratories to test the low temperature T and high baryon chemical potential region of the QCD phase diagram.

At high baryon chemical potential and low T, many QCD inspired models suggest the deconfinement transition to be a first order phase transition [1, 2]. As it is well known, first order phase transitions are triggered by the nucleation of a critical size drop of the new (stable) phase in a metastable mother phase. In the last few years, we have investigated [3]–[12] (see also Refs. [13, 14, 15, 16]) the astrophysical consequences of the nucleation process of quark matter (QM) in the core of massive pure hadronic compact stars (hadronic stars, HSs) *i.e.* neutron stars in which no fraction of QM is present. In this contribution, we report some of the main findings of these studies.

2. Equation of state of dense matter

All the results we report in the present paper are relative to the finite temperature version of the following models for the EOS of dense matter. For the hadronic phase we use the Glendenning–Moszkowski model [17, 18], and particularly the GM1 parametrization [17, 18, 9]. The nucleon coupling constants are fitted to the bulk properties of nuclear matter. The inclusion of hyperons involves new couplings, which can be written in terms of the nucleonic ones: $g_{\sigma Y} = x_\sigma\, g_\sigma$, $g_{\omega Y} = x_\omega\, g_\omega$, $g_{\rho Y} = x_\rho\, g_\rho$. In this model [17, 18] it is assumed that all the hyperons in the baryonic octet have the same coupling and, in addition, it is assumed that $x_\rho = x_\sigma$. The binding energy of the Λ particle in symmetric nuclear matter $B_\Lambda/A = -28$ MeV $= x_\omega\, g_\omega\, \omega_0 - x_\sigma\, g_\sigma \sigma$ is used [17, 18] to determine x_ω in terms of x_σ. Measured neutron star masses can be used to put additional restrictions to the possible ranges of variability of the hyperon couplings. In this work we will consider the cases $x_\sigma = 0.6$ (hereafter $\mathrm{GM1}_{0.6}$) and $x_\sigma = 0.7$ ($\mathrm{GM1}_{0.7}$). Notice that the case with $x_\sigma = 0.6$ produces stars with a larger hyperon population (for a given stellar gravitational mass) with respect to the case $x_\sigma = 0.7$ [18, 9].

For the deconfined quark phase we have used: (i) the MIT bag model EOS [19] with m_s = 150 MeV, $m_u = m_d = 0$, $\alpha_s = 0$ and different values for the bag constant B; (ii) the Nambu–Jona-Lasinio (NJL) model [20], with the lagrangian density given in Ref. [21]; (iii) the Chromo Dielectric model [22, 23].

3. Phase equilibrium

For a first-order phase transition the conditions for phase equilibrium are given by the Gibbs' phase rule

$$T_H = T_Q \equiv T\,, \qquad P_H = P_Q \equiv P_0\,, \qquad \mu_H(T, P_0) = \mu_Q(T, P_0) \tag{1}$$

where

$$\mu_H = \frac{\varepsilon_H + P_H - s_H T}{n_H}\,, \qquad \mu_Q = \frac{\varepsilon_Q + P_Q - s_Q T}{n_Q} \tag{2}$$

are the Gibbs' energies per baryon (average chemical potentials) for the hadron and quark phase respectively, ε_H (ε_Q), P_H (P_Q), s_H (s_Q) and n_H (n_Q) denote respectively the total (*i.e.*, including leptonic contributions) energy density, total pressure, total entropy density, and baryon number density for the hadron (quark) phase.

Above the "transition point" P_0 the hadronic phase is metastable, and the stable quark phase will appear as a results of a nucleation process. Virtual drops of the stable quark phase will arise from localized fluctuations in the state variables of the metastable hadronic phase. These fluctuations are characterized by a time scale $\nu_0^{-1} \sim 10^{-23}$ s. This time scale is set by the strong interactions (which are responsible for the deconfinement phase transition), and it is many orders of magnitude shorter than the typical time scale for the weak interactions. Therefore quark flavor must be conserved forming a vitual drop of QM. We will refer to this form of deconfined matter, in which the flavor content is equal to that of the β-stable hadronic system at the same pressure and temperature, as the Q*-phase. Soon afterward a critical size drop of Q*-matter is formed, the weak interactions will have enough time to act, changing the quark flavor fraction of the deconfined droplet to lower its energy, and a droplet of β-stable QM is formed (hereafter the Q-phase). This first seed of β-stable QM will trigger the conversion [24] of the pure hadronic star to a quark star (QS) (*i.e.* to a hybrid neutron star or to a strange star depending on the details of the EOS for quark matter used to model the phase transition).

The direct formation by fluctuations of a drop of β-stable QM is also possible in principle. However, it is strongly suppressed with respect to the formation of the Q*-phase drop by a factor $\sim G_{\mathrm{Fermi}}^{2N/3}$ being N the number of particles in the critical size quark drop. This is so because the formation of a β-stable drop will imply the almost simultaneous conversion of $\sim N/3$ up and down quarks into strange quarks. For a critical size β-stable drop at the center of a neutron star it is found $N \sim 100 - 1000$, and therefore the suppresion factor is actually very tiny.

In Fig. 1 we plot the Gibbs' energies per baryon for the hadron-phase and for the Q*-phase in neutrino-free matter, at different temperatures (T = 0, 10, 20, 30 MeV). Results in Fig. 1 are obtained using the GM1 model with $x_\sigma = 0.6$ for the hadronic phase and the MIT Bag model with B = 85 MeV/fm^3 for the quark phase (hereafter the $\mathrm{GM1}_{0.6}$–B85 EOS). Lines with the

Journal of Physics: Conference Series **342** (2012) 012001 doi:10.1088/1742-6596/342/1/012001

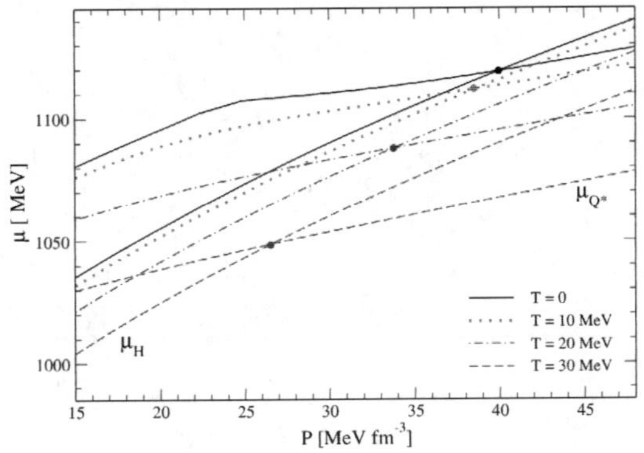

Figure 1. Gibbs energy per baryon of β-stable hadronic phase and Q*-phase, in neutrino-free matter, as a function of pressure, at different temperatures. Lines with the steeper slope refer to the hadronic phase. Full dots indicate the transition pressure P_0 for each temperature. GM1 EOS with $x_\sigma = 0.6$ for the hadronic phase and MIT bag model EOS with B = 85 MeV/fm^3 for the Q*-phase (GM1$_{0.6}$–B85 EOS).

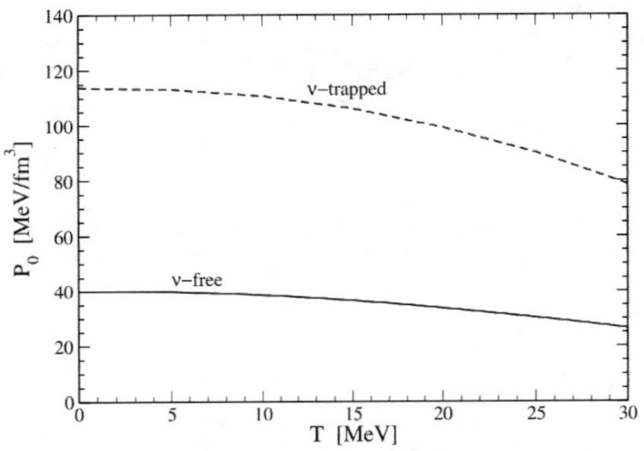

Figure 2. Phase equilibrium curve between the β-stable hadronic phase and the Q* phase. The continuous curve is relative to neutrino-free matter, the dashed curve to matter with trapped neutrinos. EOS: GM1 with $x_\sigma = 0.6$ plus MIT Bag model with B = 85 MeV/fm^3.

steeper slope refer to the hadron phase. As we see, the transition pressure P_0 (indicated by a full dot) decreases when the hadronic matter temperature is increased.

The phase equilibrium curve $P_0(T)$ between the β-stable hadronic phase and the Q^*-phase is shown in Fig. 2 for neutrino-free matter and matter with trapped neutrinos, making use of the GM1$_{0.6}$–B85 EOS. The region of the P_0–T plane above each curve represents the deconfined Q*-phase. As expected [25, 26, 6, 27] neutrino trapping in β-stable hadronic matter inhibits the quark deconfinement phase transition, thus the global effect of neutrino-trapping is to produce a shift of the phase equilibrium curve toward higher values of the pressure in the P_0–T plane.

In Fig. 3, we show the phase equilibrium curve, for neutrino-free matter, in the case of the NJL model (left panel) or the Chromo Dielectric model (right panel) to describe the deconfined phase. For the hadron phase we take the GM1 model with $x_\sigma = 0.7$ (GM1$_{0.7}$) in both cases. Notice that in the case of the NJL model the transition pressure $P_0(T)$ is substantially higher than the one in the case of the MIT Bag or Chromo Dielectric models to describe the deconfined phase. As discussed in detail in Ref. [12], this behaviour can be traced back the large value of the stange quark effective mass in the NJL model.

As is well known, for a first-order phase transition the derivative dP_0/dT is related to the specific latent heat $\mathcal{Q}$ of the phase transition by the Clapeyron-Clausius equation

$$\frac{dP_0}{dT} = -\frac{n_H n_{Q^*}}{n_{Q^*} - n_H}\frac{\mathcal{Q}}{T} \tag{3}$$

$$\mathcal{Q} = \tilde{W}_{Q^*} - \tilde{W}_H = T(\tilde{S}_{Q^*} - \tilde{S}_H) \tag{4}$$

where $\tilde{W}_H$ ($\tilde{W}_{Q^*}$) and $\tilde{S}_H$ ($\tilde{S}_{Q^*}$) denote the enthalpy per baryon and entropy per baryon for the hadron (quark) phase, respectively. The specific latent heat $\mathcal{Q}$ and the phase numbers densities

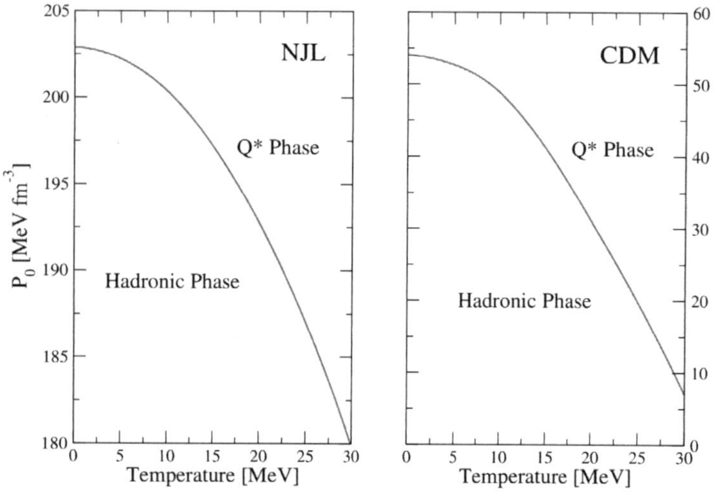

Figure 3. Phase equilibrium curve between the β-stable hadronic phase and the Q*-phase. The EOS for the hadronic phase is obtained using the GM1 model with $x_\sigma = 0.7$, the EOS for the Q^*-phase using the NJL model (left panel) or the Chromo Dielectric model (right panel).

n_H and n_{Q^*} at phase equilibrium are reported in Tables 1 and 2 in the case of the $\mathrm{GM1}_{0.6}$–B85 equation of state. Results in Table 1 refer to neutrino-free matter, whereas those in Table 2 refer to matter with trapped neutrinos. As expected for a first-order phase transition, one has a discontinuity jump in the phase number densities: in our particular case $n_{Q^*}(T, P_0) > n_H(T, P_0)$. This result, together with the positive value of $\mathcal{Q}$ (*i.e.* the deconfinement phase transition absorbs heat), tells us (see Eq. (3)) that the phase transition temperature decreases with pressure (as in the melting of ice).

The effect of neutrino trapping on the phase equilibrium properties of the system can be seen comparing the results reported in Tables 1 and 2. As we see, the phase number densities n_H and n_{Q^*} at phase equilibrium are shifted to higher values, and the specific latent heat $\mathcal{Q}$ is increased with respect to the neutrino-free matter case.

4. Quark matter nucleation in cold hadronic stars

In a cold ($T = 0$) and neutrino-free pure hadronic star the formation of the first drop of QM could take place solely via a quantum nucleation process. The basic quantity needed to calculate the nucleation time is the energy barrier separating the Q*-phase from the metastable hadronic phase. This energy barrier, which represents the difference in the free energy of the system with and without a Q*-matter droplet, can be written as [28, 29]

$$U(\mathcal{R}) = \frac{4}{3}\pi n_{Q^*}(\mu_{Q^*} - \mu_H)\mathcal{R}^3 + 4\pi\sigma\mathcal{R}^2 \qquad (5)$$

where $\mathcal{R}$ is the radius of the droplet (supposed to be spherical), and σ is the surface tension for the surface separating the hadron from the Q*-phase. The energy barrier has a maximum at the critical radius $\mathcal{R}_c = 2\sigma/[n_{Q^*}(\mu_H - \mu_{Q^*})]$. We neglected the term associated with the curvature energy and also the terms connected with the electrostatic energy, since they are known to only introduce small corrections [29, 5]. The value of the surface tension σ for the interface separating the quark and hadron phase is poorly known, and typically values used in the literature range within 10 – 50 MeV fm^{-2} [30]). We assume σ to be temperature independent and we take $\sigma = 30$ MeV fm^{-2}.

The quantum nucleation time τ_q can be straightforwardly evaluated within a semi-classical approach [28, 29, 4] and it can be expressed as

$$\tau_q = (\nu_0 p_0 N_c)^{-1} \ , \qquad (6)$$

Table 1. The specific (*i.e.* per baryon) latent heat Q and the phase number densities n_H and n_{Q^*} at phase equilibrium. GM1 EOS with $x_\sigma = 0.6$ for the hadronic phase, MIT bag model with $B = 85$ MeV/fm^3 for the quark phase. Results for neutrino-free matter.

T MeV	Q MeV	n_{Q^*} fm^{-3}	n_H fm^{-3}	P_0 MeV/fm^3
0	0.00	0.453	0.366	39.95
5	0.56	0.451	0.364	39.74
10	2.40	0.447	0.358	38.58
15	5.71	0.439	0.348	36.55
20	10.60	0.428	0.334	33.77
25	17.17	0.414	0.316	30.36
30	25.44	0.398	0.294	26.53

Table 2. Same as Table 1, but with trapped neutrinos.

T MeV	Q MeV	n_{Q^*} fm^{-3}	n_H fm^{-3}	P_0 MeV/fm^3
0	0.00	0.603	0.516	113.77
5	0.65	0.601	0.514	113.11
10	2.87	0.594	0.509	110.69
15	6.78	0.580	0.499	106.17
20	12.65	0.560	0.483	99.18
25	20.21	0.534	0.462	90.12
30	29.88	0.502	0.434	78.65

where p_0 is the probability of tunneling the energy barrier $U(\mathcal{R})$ in its ground state, ν_0 is the oscillation frequency of a virtual drop of the Q*-phase in the potential well, and $N_c \sim 10^{48}$ is the number of nucleation centers expected in the innermost part ($r \leq R_{nuc} \sim 100$ m) of the HS, where the pressure and temperature (when we will consider the finite T case) can be considered constant and equal to their central values.

Thus a pure hadronic star, having a central pressure P_c larger than the static transition pressure P_0 for the formation of the Q*-phase, is metastable to the "decay" (conversion) to a quark star (QS) *i.e.* to a more compact stellar configuration in which deconfined quark matter is present. These metastable HSs have a *mean-life time* which is related to the nucleation time to form the first critical-size drop of deconfined matter in their interior (the actual *mean-life time* of the HS will depend on the mass accretion or on the spin-down rate which modifies the nucleation time via an explicit time dependence of the stellar central pressure). Following Refs. [3, 4, 5] we define as *critical mass* M_{cr} of the metastable HSs, the value of the gravitational mass for which the nucleation time is equal to one year: $M_{cr} \equiv M_{HS}(\tau_q = 1\text{yr})$. Pure hadronic stars with $M_{HS} > M_{cr}$ are very unlikely to be observed. Thus M_{cr} plays the role of an *effective maximum mass* for the hadronic branch of compact stars. Notice that the Oppenheimer–Volkov maximum mass M_{HS}^{max} is determined by the overall stiffness of the EOS for hadronic matter, whereas the value of M_{cr} will depend in addition on the bulk properties of the EOS for quark matter and on the properties at the interface between the confined and deconfined phases of

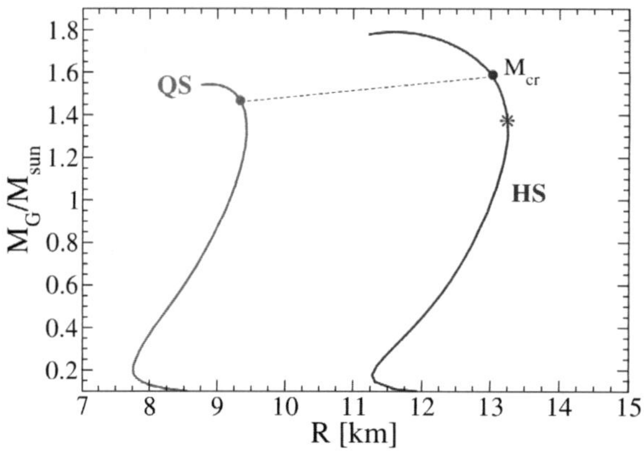

Figure 4. Mass-radius relation for a pure hadronic stars (HS) and hybrid star configurations (QS). The configuration marked with an asterisk represents the HS for which $\tau_q = \infty$ (*i.e.* $P_c = P_0$). The conversion process of the HS, with a gravitational mass equal to M_{cr}, into the final QS is denoted by the full circles connected by a dashed line. Results are relative to the GM1 model with $x_\sigma = 0.6$ for the hadronic phase and the MIT Bag model EOS with B = 85 MeV/fm^3 for the quark phase. The surface tension is $\sigma = 30$ MeV/fm^2. Stellar masses are in unit of the mass of the sun, $M_{sun} = 1.989 \times 10^{33}$ g.

matter (*e.g.*, the surface tension σ).

These findings are exemplified in Fig. 4, where we show the mass-radius (MR) curve for hadronic stars (HS) and that for quark stars (QS). The configuration marked with an asterisk on the hadronic MR curve represents the HS for which the central pressure is equal to P_0 and thus $\tau_q = \infty$. The full circle on the HS sequence represents the critical mass configuration M_{cr}, in the case $\sigma = 30$ MeV/fm^2. The full circle on the QS mass-radius curve represents the hybrid star which is formed from the conversion of the hadronic star with $M_{HS} = M_{cr}$. We assume [24] that during the stellar conversion process the total number of baryons in the star is conserved. The stellar conversion process, described so far, will start to populate the new branch of quark stars (the part of the QS sequence above the full circle). Long term accretion on the QS can next produce stars with masses up to the limiting mass M_{QS}^{max} for the quark star configurations.

5. Quark matter nucleation in proto-hadronic stars

A neutron star at birth (proto-neutron star) is very hot (T = 10 – 30 MeV) with neutrinos being still trapped in the stellar interior [31, 32, 25, 33, 34, 35]. Subsequent neutrino diffusion causes deleptonization and heats the stellar matter to an approximately uniform entropy per baryon $\tilde{S} = 1 - 2$ (in units of the Boltzmann's constant k_B). Depending on the stellar composition, during this stage neutrino escape can lead the more "massive" stellar configurations to the formation of a black hole [36, 25]. However, if the mass of the star is sufficiently small, the star will remain stable and it will cool to temperatures well below 1 MeV within a cooling time $t_{cool} \sim$ a few 10^2 s, as the neutrinos continue to carry energy away from the stellar material [31, 25, 33]. Thus in a proto-neutron star, the quark deconfinement phase transition will be likely triggered by a thermal nucleation process [37, 38, 39, 40]. In fact, for sufficiently high temperatures, thermal nucleation is a much more efficient process with respect to the quantum nucleation mechanism. In the following we will establish the physical conditions under which a newborn hadronic star (proto-hadronic star, PHS) could survive the early stages of its evolution without "decaying" to a quark star.

Journal of Physics: Conference Series **342** (2012) 012001 doi:10.1088/1742-6596/342/1/012001

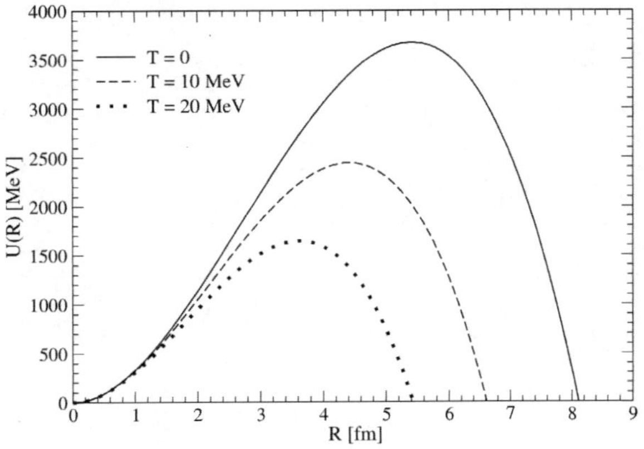

Figure 5. Energy barrier for a virtual drop of the Q*-phase in β-stable neutrino-free hadronic matter as a function of the droplet radius and for different temperatures for a fixed pressure $P = 57$ MeV/fm^3. Results are relative to the GM1$_{0.6}$–B85 equation of state and surface tension $\sigma = 30$ MeV/fm^2.

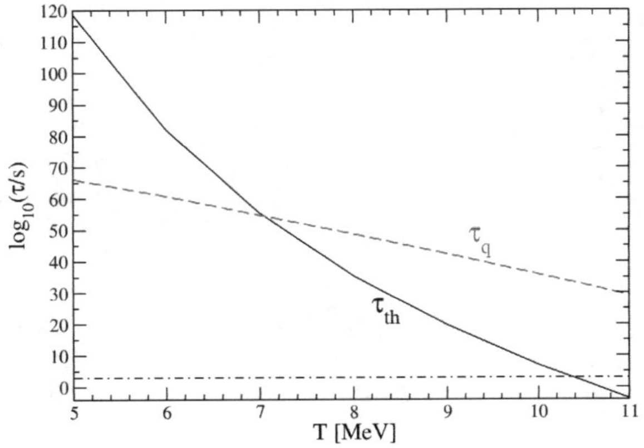

Figure 6. Thermal (τ_{th}) and quantum (τ_q) nucleation time of quark matter (Q*-phase) in β-stable neutrino-free hadronic matter as a function of temperature at fixed pressure $P = 57$ MeV/fm^3. The crossover temperature is $T_{co} = 7.05$ MeV. The limiting conversion temperature for the proto-hadronic star is, in this case, $\Theta = 10.3$ MeV, obtained from the intersection of the thermal nucleation time curve (continuous line) and the dot-dashed line representing $\log_{10}(\tau/\mathrm{s}) = 3$. The surface tension is $\sigma = 30$ MeV/fm^2. Results are relative to GM1$_{0.6}$–B85 EOS.

According to the Langer theory [41, 42] of homogeneous nucleation the thermal nucleation rate can be written [42] as

$$I = \frac{\kappa}{2\pi}\Omega_0 \exp(-U(\mathcal{R}_c, T)/T) \tag{7}$$

where κ is the so-called dynamical prefactor, which is related to the growth rate of the drop radius $\mathcal{R}$ near the critical radius ($\mathcal{R}_c$), and Ω_0 is the so-called statistical prefactor, which measures the phase-space volume of the saddle-point region around $\mathcal{R}_c$. We have used [10, 11] for κ and Ω_0 the expressions derived in Refs. [43, 44], where the Langer nucleation theory has been extended to the case of first order phase transitions occurring in relativistic systems, as in the case of the quark

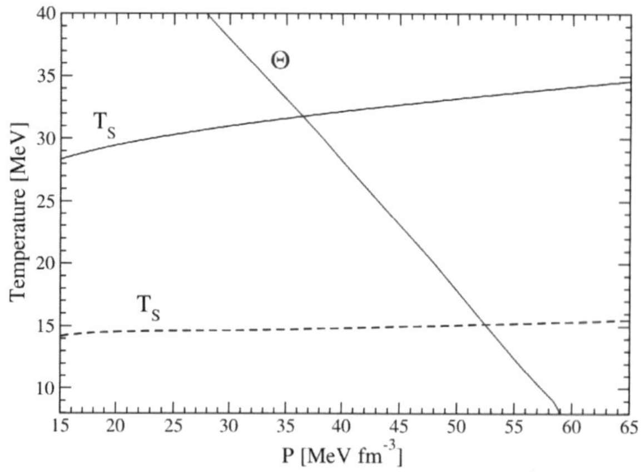

Figure 7. The limiting conversion temperature Θ for a newborn hadronic star as a function of the central stellar pressure. Newborn hadronic stars with a central temperature and pressure located on the right side of the curve $\Theta(P)$ will nucleate a Q*-matter drop during the early stages of their evolution, and will finally evolve to cold and deleptonized quark stars, or will collapse to black holes. The lines labeled T_S represent the stellar matter temperature as a function of pressure at fixed entropies per baryon $\tilde{S}/k_B = 1$ (dashed line) and 2 (solid line). Results are relative to GM1$_{0.6}$–B85 EOS.

deconfinement transition. The dominant factor in the nucleation rate (7) is the exponential, in which $U(\mathcal{R}_c, T)$ is the activation energy, *i.e.* the change in the free energy of the system required to activate the formation of a critical size droplet. This quantity has been calculated [10, 11] using the generalization of Eq. (5) to the case of $T \neq 0$ (thus using finite temperature EOS for the hadronic and the quark phases), and assuming a temperature independent surface tension.

The thermal nucleation time τ_{th}, relative to the innermost stellar region ($V_{nuc} = (4\pi/3)R^3_{nuc}$, with $R_{nuc} \sim 100$ m) where almost constant pressure and temperature occur, can thus be written as

$$\tau_{th} = (V_{nuc}\, I)^{-1} \,. \qquad (8)$$

In Fig. 5, we represent the energy barrier for a virtual drop of the Q*-phase in the neutrino-free hadronic phase as a function of the droplet radius and for different temperatures at a fixed pressure $P = 57$ MeV/fm^3. Results in Fig. 5 are obtained using the GM1$_{0.6}$–B85 equation of state. As expected, from the results plotted in Fig. 1, the energy barrier $U(\mathcal{R}, T)$ and the droplet critical radius $\mathcal{R}_c$ decrease as the matter temperature is increased. This effect favors the Q*-phase formation and, in particular, increases (decreases) the quantum nucleation rate (nucleation time τ_q) with respect to the corresponding quantities calculated at $T = 0$.

In Fig. 6, we plot the quantum and thermal nucleation times of the Q*-phase in β-stable neutrino-free hadronic matter as a function of temperature and at a fixed pressure $P = 57$ MeV/fm^3. As expected, we find a crossover temperature T_{co} above which thermal nucleation is dominant with respect to the quantum nucleation mechanism. For the case reported in Fig. 6, we have $T_{co} = 7.05$ MeV and the corresponding nucleation time is $\log_{10}(\tau/\mathrm{s}) = 54.4$.

Having in mind the physical conditions in the interior of a PHS [31, 25], to establish if this star will survive the early stages of its evolution without decaying to a quark star, one has to compare the quark matter nucleation time $\tau = \min(\tau_q, \tau_{th})$ with the cooling time $t_{cool} \sim$ a few 10^2 s. If $\tau >> t_{cool}$ then quark matter nucleation will not likely occur in the newly formed star, and this star will evolve to a cold deleptonized configuration. We thus introduce the concept of *limiting conversion temperature* Θ for the proto-hadronic star and define it as the value of the stellar

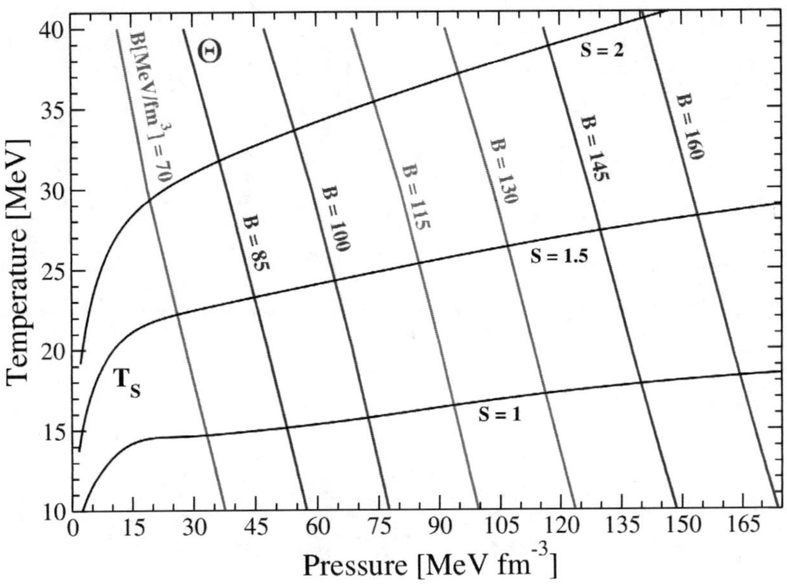

Figure 8. The limiting conversion temperature Θ for a newborn hadronic star as a function of the central stellar pressure for different values of the bag constant B. The lines labeled T_S represent the stellar matter temperature as a function of pressure at fixed entropies per baryon $\tilde{S}/k_B = 1$, 1.5, 2. Results for neutrino-free matter.

central temperature T_c for which the Q*-matter nucleation time is equal to 10^3 s. The limiting conversion temperature Θ will clearly depend on the value of the stellar central pressure (and thus on the value of the stellar mass).

The limiting conversion temperature Θ is plotted in Fig. 7 as a function of the stellar central pressure. A proto-hadronic star with a central temperature $T_c > \Theta$ will likely nucleate a Q*-matter drop during the early stages of its evolution, and will finally evolve to a cold and deleptonized quark star, or will collapse to a black hole (depending on the value of the stellar baryonic mass M_B and on the EOS).

In Fig. 8 we plot the limiting conversion temperature Θ for a newborn HS for different values of the bag constant. The increase in B produces a growth of the region of the P-T plane where the proto-hadronic star could survive Q* nucleation and thus evolve to a cold hadronic star.

For an isoentropic stellar core [31, 25], the central temperature of the proto-hadronic star is given, for the GM1$_{0.6}$ EOS model, by the lines labeled by T_S in Fig. 7, relative to the case $\tilde{S} = 1\ k_B$ (dashed curve) and $\tilde{S} = 2\ k_B$ (continuous curve). The intersection point (P_S, Θ_S) between the two curves $\Theta(P)$ and $T_S(P)$ thus gives the central pressure and temperature of the configuration that we denote as the *critical mass* configuration of the proto-hadronic stellar sequence. Taking $\sigma = 30$ MeV/fm^2 and $\tilde{S} = 2\ k_B$ we get $M_{cr} = 1.390\ M_{sun}$ for the gravitational critical mass and $M_{B,cr} = 1.492\ M_{sun}$ for the baryonic critical mass (being $M_{sun} = 1.989 \times 10^{33}$ g the mass of the sun).

The evolution of a PHS within our scenario is delineated in Fig. 9, where we plot the appropriate stellar equilibrium sequences in the gravitational–baryonic mass plane for the GM1$_{0.6}$–B85 EOS and $\sigma = 30$ MeV/fm^2. The upper (red) line represents the PHS sequence, *i.e.* isoentropic HSs ($\tilde{S} = 2\ k_B$) and neutrino-free matter. The middle (blue) line represents the cold HS sequence. The asterisk and the full circle on these lines represent respectively the stellar configuration with $\tau = \infty$ and the critical mass configuration. Finally, the lower (green)

Journal of Physics: Conference Series **342** (2012) 012001 doi:10.1088/1742-6596/342/1/012001

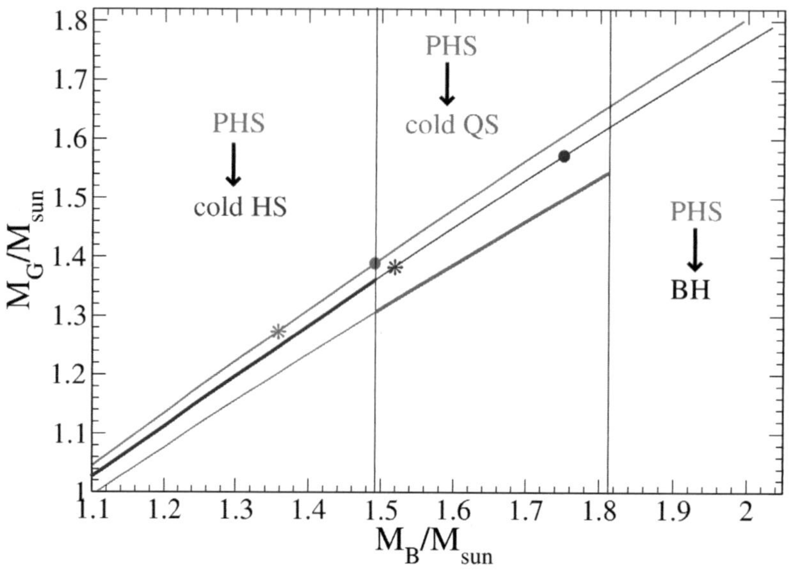

Figure 9. Stellar equilibrium sequences in the gravitational–baryonic mass plane for the GM1$_{0.6}$–B85 EOS and σ = 30 MeV/fm^2. The upper (red) line represents the PHS sequence ($\tilde{S} = 2\ k_B$). The middle (blue) line represents the cold HS sequence. The asterisk and the full circle on these lines represent respectively the stellar configuration with $\tau = \infty$ and the critical mass configuration M_{cr}. The lower (green) line represent the cold QS sequence. Assuming $M_B =$ const, the evolution of a PHS in this plane occurs along a vertical line.

line represent the cold QS sequence. We assume $M_B =$ const during these stages of the stellar evolution. Thus according to the results in Fig. 9, proto-hadronic stars with a baryonic mass $M_B < 1.492\ M_{sun}$ will survive Q*-matter *early nucleation* (*i.e.* nucleation within the cooling time $t_{cool} \sim$ a few 10^2 s) and in the end they will form stable ($\tau = \infty$) cold hadronic stars. Proto-hadronic stars with $1.492\ M_{sun} \leq M_B < 1.813\ M_{sun}$ (the maximum baryonic mass $M^{max}_{B,QS}$ of the cold QS sequence for the present EOS) will experience early nucleation of a Q*-matter drop and will ultimately form a cold deleptonized quark star. The last possibility is for PHSs having $M_B > 1.813\ M_{sun}$. In this case the early nucleation of a Q*-matter drop will trigger a stellar conversion process to a cold QS configuation with $M_B > M^{max}_{B,QS}$, thus these PHSs will finally forms black holes.

The outcomes of this scenario are not altered by neutrino trapping effects in hot β-stable hadronic matter. In fact, for proto-hadronic stars with trapped neutrinos (νPHSs) (with a lepton fraction $Y_L = 0.4$ and $\tilde{S} = 2\ k_B$) we find [11] (for the same EOS and surface tension σ used in Fig. 9) a critical baryonic mass $M_{B,cr} = 1.96\ M_{sun}$. Thus νPHSs with $M_B \geq M_{B,cr}$ after neutrino escape and cooling will finally evolve to black holes. The fate of a νPHS with $M_B < M_{B,cr}$ is the same as the corresponding neutrino-free PHS with equal baryonic mass.

In Fig. 10 we plot the PHS, cold HS, and cold QS sequences in the gravitational–baryonic mass plane for the case of the NJL model for the quark phase and the GM1 model in the case of pure nucleonic matter (right panel) or hyperonic matter with $x_\sigma = 0.7$ (left panel). It is clearly seen that in the case of the NJL model it is almost impossible to populate the QS branch. Cold quark stars can be formed in the case of $x_\sigma = 0.7$ (left panel) for a very narrow range of baryonic stellar masses $2.20 < M_B/M_\odot < 2.23 M_\odot$.

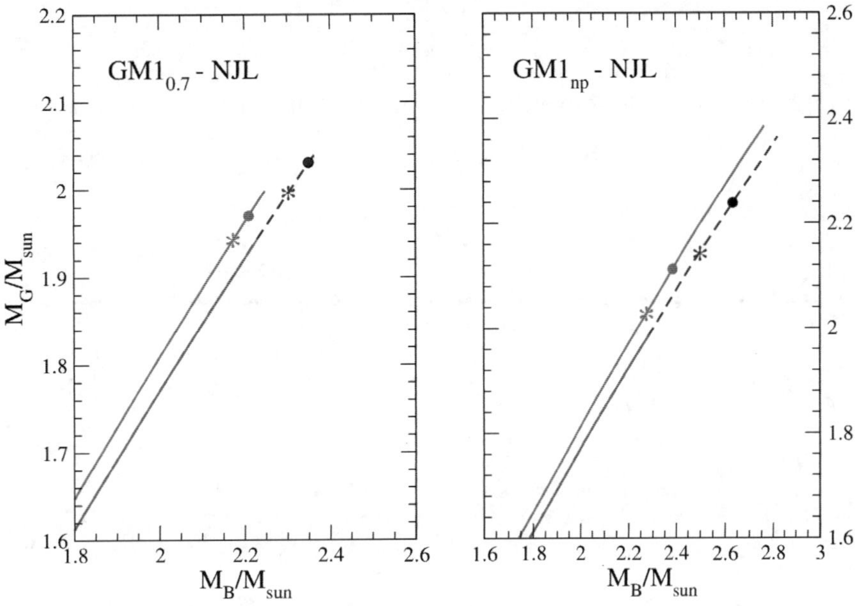

Figure 10. Same as in the previous figure but in the case of the NJL model for the quark phase and the Glendenning–Moszkowski model for hyperonic matter with $x_\sigma = 0.7$ ($GM1_{0.7}$) (left panel) and pure nucleonic matter $GM1_{np}$ (right panel).

6. Conclusions

In summary, in this work we have studied the quark deconfinement phase transition in hot β-stable hadronic matter and explored some of its consequences for the physics of neutron stars at birth. We calculated and compared the nucleation time due to thermal and quantum nucleation mechanisms, and computed the crossover temperature above which thermal nucleation dominates the finite temperature quantum nucleation mechanism. In addition, we introduced the new concept of *limiting conversion temperature* Θ for proto-hadronic stars and extended the concept of critical mass (introduced in Refs. [3, 4, 5]) to the case of finite temperature hadronic stars.

Our main finding is that proto-hadronic stars with a gravitational mass lower than the critical mass M_{cr} could survive the early stages of their evolution without decaying to a quark star. This outcome contrasts with the predictions of earlier studies [37, 38, 39, 40] where it was inferred that all the pure hadronic compact stars, with a central temperature above $2-3$ MeV, are converted to quark stars within the first seconds after their birth. However, the prompt formation of a critical size drop of quark matter could take place when $M > M_{cr}$. These proto-hadronic stars evolve to cold and deleptonized quark stars or collapse to a black holes.

We found that for the NJL model the presence of hyperons disfavour the phase transition pushing the transition point to very high densities. In addition, in the case of the NJL model it is almost impossible to populate the quark star branch and quark matter nucleation will lead to the formation of a black hole. Thus within the NJL model for the quark phase, all compact stars are pure hadronic stars.

Accurate measurements of both the mass and radius of a few individual "neutron stars" [45, 46] or measurements of temperature profiles of accretion discs around rapidly spinning compact stars [47] could shed light on the validity of the scenario discussed in the present work.

Journal of Physics: Conference Series **342** (2012) 012001 doi:10.1088/1742-6596/342/1/012001

References

[1] Hsu S D H and Schwetz M 1998 *Phys. Lett.* **B 432** 2003
[2] Fodor Z and Katz S D 2004 *JHEP* **04** 050
[3] Berezhiani Z, Bombaci I, Drago A, Frontera F and Lavagno A 2002 *Nucl. Phys. B - Proceedings Supplement* **113** 268
[4] Berezhiani Z, Bombaci I, Drago A, Frontera F and Lavagno A 2003 *Astrophys. Jour.* **586** 1250
[5] Bombaci I, Parenti I and Vidaña I 2004 *Astrophys. Jour.* **614**, 314
[6] Vidaña I, Bombaci I and Parenti I 2005 *J. Phys. G: Nucl. Part. Phys.* **31** S1165
[7] Lugones G and Bombaci I 2005 *Phys. Rev.* D **72** 065021
[8] Bombaci I, Lugones G and Vidaña I 2007 *Astron. and Astrophys.* **462** 1017
[9] Bombaci I, Panda P K, Providência C and Vidaña I 2008 *Phys. Rev.* D **77** 083002
[10] Bombaci I, Logoteta D, Panda P K, Providência C and Vidaña I 2009 *Phys. Lett.* B **680** 448
[11] Bombaci I, Logoteta D, Providência C and Vidaña I 2011 *Astron. and Astrophys.* **528** A71
[12] Logoteta D, Bombaci I, Providência C and Vidaña I 2011 *Phys. Rev.* D **XX** ppp (submitted)
[13] Drago A, Lavagno A and Pagliara G 2004 *Phys. Rev.* D **69** 057505
[14] Drago A, Pagliara G and Schaffner-Bielich J 2008 *J. Phys. G* **35** 014052
[15] Bambi C and Drago A 2008 *Astropart. Phys.* **29** 223
[16] Drago A, Lavagno A and Parenti I 2007 *Astrophys. Jour.* **659** 1519
[17] Glendenning N K and Moszkowski S A 1991 *Phys. Rev. Lett.* **67** 2414
[18] Glendenning N K 2000, Compact Stars: Nuclear Physics, Particle Physics, and General Relativity, Springer, New York.
[19] Farhi E and Jaffe R L 1984 *Phys. Rev.* D **30** 2379
[20] Nambu Y and Jona-Lasinio G 1961 *Phys Rev.* **122** 345
[21] Rehberg P, Klevansky S P and Hufner J 1996 *Phys. Rev.* C **53** 410
[22] Pirner H J, Chanfray G and Nachtmann O 1984 *Phys. Lett.* B **147** 249
[23] Drago A and Tambini E 1999 *J. Phys.* G **25** 971
[24] Bombaci I and Datta B 2000 *Astrophys. Jour.* **530** L69
[25] Prakash M, Bombaci I, Prakash M, Ellis P J, Lattimer J M and Knorren R 1997 Phys. Rep. **280** 1
[26] Lugones G and Benvenuto O G 1998 *Phys. Rev.* D **58** 083001
[27] Lugones G, Grunferld A G, Scoccola N N and Villavicencio C 2009 *Phys. Rev.* D **80** 045017
[28] Lifshitz I M and Kagan Y 1972 *Sov. Phys. JETP* **35** 206
[29] Iida K and Sato K 1998 *Phys. Rev.* C **58** 2538
[30] Heiselberg H, Pethick C J and Staubo E F 1993 *Phys. Rev. Lett.* **70** 1355
[31] Burrow A and Lattimer J M 1986 *Astrophys. Jour.* **307** 178
[32] Bombaci I, Prakash M, Prakash M, Ellis P J, Lattimer J M and Brown G E 1995 *Nucl. Phys.* A **583** C623
[33] Pons J A, Reddy S, Prakash M, Lattimer J M and Miralles J A 1999 *Astrophys. Jour.* **513** 780
[34] Vidaña I, Bombaci I, Polls A and Ramos A 2003 *Astron. and Astrophys.* **399** 687
[35] Margueron J, Vidaña I and Bombaci I 2003 *Phys. Rev.* C **68** 055806
[36] Bombaci I 1996 *Astron. and Astrophys.* **305** 871
[37] Horvath J E, Benvenuto O G and Vucetich H 1992 *Phys. Rev.* D **45** 3865
[38] Horvath J E 1994 *Phys. Rev.* D **49** 5590
[39] Olesen M L and Madsen J 1994 *Phys. Rev.* D **49** 2698
[40] Harko T, Cheng K S and Tang P S 2004 *Astrophys. Jour.* **608** 945
[41] Langer J S 1969 *Ann. Phys. (N.Y.)* **54** 258
[42] Langer J S and Turski L A 1973 *Phys. Rev.* A **8** 3230
[43] Csernai L and Kapusta J I 1992 *Phys. Rev.* D **46** 1379
[44] Venugopalan R and Vischer A P 1994 *Phys. Rev.* E **49** 5849
[45] Bhattacharyya S 2010 *Adv. Space Res.* **45** 949
[46] Steiner A W, Lattimer J M and Brown E F 2010 *Astrophys. Jour.* **722** 33
[47] Bhattacharyya S, Thampan A V and Bombaci I 2001 *Astr. and Astrophys.* **372** 925

Journal of Physics: Conference Series **342** (2012) 012002 doi:10.1088/1742-6596/342/1/012002

The Symmetry energy of nuclear matter under a strong magnetic field

R. Casali[1a], C. Providência[2b], D. Menezes[1c]

[1] Departamento de Fisica (CFM), Universidade Federal de Santa Catarina, Florianópolis (SC) caixa postal 476, CEP 88.040-900, Brazil

[2] Centro de Fisica Computacional, University of Coimbra, 3004-516, Coimbra, Portugal

E-mail: [a]`rcasali@fisica.ufsc.br`,[b]`cp@teor.fis.uc.pt`,[c]`debora@fsc.ufsc.br`

Abstract. In this work we we study the effects caused by strong magnetic fields on the symmetry energy and its slope. We compare two models under three intensities of magnetic field, we investigate the behaviour for the conditions with and without anomalous magnetic moment. It is discussed the possible effects on the structure of the pasta phase.

1. Introduction

Asymmetric nuclear matter [1] is an important issue for both theoretical and experimental studies, specially for those interested in the description of stellar matter of compact stars. With the advent of new radioactive beams which will be operating in the near future, this area appears promising.

The symmetry energy is a quantity related to the neutron-proton asymmetry in the equation of state of nuclear matter, it is an important subject both for nuclear physics and astrophysics and is extremely important to the understanding of neutron star structure and composition [2]. In particular, the symmetry energy slope defines nuclear matter properties, such as, neutron skin, neutron dripline, binding energy and core-crust transition density [3]. We want to study the effect of the magnetic field on the symmetry energy.

We consider a system formed by protons and neutrons, interacting via the exchange of σ-ω-ρ mesons, in the presence of an uniform magnetic field B along the z-axis. Three possible values of magnetic fields, $B = 0, 10^{17}$ and $10^{18}G$ are taken. We will also study the effect of the anomalous magnetic moment for protons and neutrons. The NL3 and FSU parameterizations are used to describe the equations of state in the relativistic nuclear mean field formalism. These models present very different density dependence of the symmetry energy. At saturation, the slope L is, respectively, 118 and 60 MeV.

2. Formalism

The Lagrangian used is given by:

$$\begin{aligned}\mathcal{L} = \quad & \overline{\psi}_b[i\gamma_\mu\partial^\mu + q_b\gamma_\mu A^\mu - g_\rho\tau_{3b}\gamma_\mu\vec{\rho}^\mu - m_b^* - g_\omega\gamma_\mu\omega^\mu - k_b\sigma_{\mu\nu}F^{\mu\nu}]\psi_b \qquad (1)\\ & +\frac{1}{2}(\partial_\mu\sigma\partial^\mu\sigma - m_\sigma^2\sigma^2) + \frac{1}{3}bm_n(g_\sigma\sigma)^3 - \frac{1}{4}c(g_\sigma\sigma)^4 + \frac{1}{2}m_\omega^2\omega_\mu\omega^\mu - \frac{1}{4}\Omega_{\mu\nu}\Omega^{\mu\nu}\end{aligned}$$

Journal of Physics: Conference Series **342** (2012) 012002 doi:10.1088/1742-6596/342/1/012002

$$+\frac{1}{2}m_\rho^2\vec{\rho}_\mu\cdot\vec{\rho}_\mu-\frac{1}{4}F^{\mu\nu}F_{\mu\nu}-\frac{1}{4}P^{\mu\nu}P_{\mu\nu}+\frac{1}{4!}\xi g_\omega^4(\omega_\mu\omega^\mu)^2+\Lambda_\omega(g_\rho^2\vec{\rho}_\mu\cdot\vec{\rho}^\mu)(g_\omega^2\omega_\mu\omega^\mu).$$

The Lagrangian presents an isodoublet nucleon field (ψ) interacting via the exchange of the scalar meson (σ), the vector meson (ω^μ), the isovector meson (ρ^μ) and the photon (A^μ), where $m_b^* = (m_b - g_\sigma\sigma)$, $\Omega_{\mu\nu} = \partial_\mu\omega_\nu - \partial_\nu\omega_\mu$, $P_{\mu\nu} = \partial_\mu\rho_\nu - \partial_\nu\rho_\mu$ and $F_{\mu\nu} = \partial_\mu A_\nu - \partial_\nu A_\mu$. The nucleon anomalous magnetic moments are introduced via the coupling of the baryons to the electromagnetic field tensor $\sigma_{\mu\nu} = \frac{i}{2}[\gamma_\mu,\gamma_\nu]$ and strength $k_p = 1.79285$ for the proton and $k_n = -1.91315$ for the neutron [4, 5]. The parameters ξ, used to describe the omega-meson self-interactions, and Λ_ω, that modifies the density-dependence of the symmetry energy, are included so we can compare the two models employed [6, 7].

Using the Lagrangian (1) and the mean-field approximation, the meson field expressions follow from the Euler-Lagrange equations:

$$\begin{aligned} m_\sigma^2\sigma_0 &= -cg_\sigma^4\sigma_0^3 - bm_n g_\sigma^3\sigma_0^2 + g_\sigma n^s \\ m_\omega^{*2}\omega_0 &= -\frac{\xi}{6}g_\omega^4\omega_0^3 + g_\omega n_b \\ m_\rho^{*2}\rho_{03} &= \tau_{3b} g_\rho n_3, \end{aligned} \tag{2}$$

with $m_\rho^{*2} = m_\rho^2 + 2\Lambda_\omega g_\rho^2 g_\omega^2\omega_0^2$, $m_\omega^{*2} = m_\omega^2 + 2\Lambda_\omega g_\rho^2 g_\omega^2\rho_{03}^2$, $n^s = n_p^s + n_n^s$, $n_b = n_p + n_n$ and $n_3 = n_p - n_n$. Where:

$$\begin{aligned} n_p^s &= \frac{q_pBm_p^*}{2\pi^2}\sum_\nu^{\nu_{max}}\sum_s\frac{\sqrt{m_p^{*2}+2\nu q_pB}-s\mu_Nk_pB}{\sqrt{m_p^{*2}+2\nu q_pB}}\ln\left|\frac{p_{F,\nu,s}^p+E_F^p}{\sqrt{m_p^{*2}+2\nu q_pB}-s\mu_Nk_pB}\right| \\ n_p^s &= \frac{m_n^*}{4\pi^2}\sum_s\left[E_F^np_{F,s}^n-\overline{m}_n^2\ln\left|\frac{p_{F,s}^n+E_F^n}{\overline{m}_n}\right|\right] \\ n_p &= \frac{q_pB}{2\pi^2}\sum_\nu^{\nu_{max}}\sum_s p_{F,\nu,s}^p \\ n_n &= \frac{1}{2\pi^2}\sum_s\left[\frac{1}{3}(p_{F,s}^n)^3-\frac{1}{2}s\mu_Nk_nB\left(\overline{m}_np_{F,s}^n+E_F^{n2}\left(\arcsin\left(\frac{\overline{m}_n}{E_F^n}\right)-\frac{\pi}{2}\right)\right)\right]. \end{aligned} \tag{3}$$

Table 1. Parameter sets for the two models discussed in the text. The meson masses m_σ, m_ω and m_ρ are given in MeV. The nucleon mass is 939 MeV in both models.

Model	m_σ	m_ω	m_ρ	g_σ	g_ω	g_ρ	c	b	ξ	Λ_ω
NL3	508.194	782.501	763.000	10.217	12.868	8.948	-0.002651	0.002055	0.00	0.00
FSU	491.5	782.5	763.0	10.592	14.302	11.767	0.0396	0.000756	0.06	0.03

The energy spectra for protons and neutrons are given by [8]:

$$\begin{aligned} E_{\nu,s}^p &= \sqrt{p_z^2+(\sqrt{m_p^*+2\nu q_pB}-s\mu_Nk_pB)^2}+g_\omega\omega^0+\frac{1}{2}g_\rho\rho^0 \\ E_s^n &= \sqrt{p_z^2+(\sqrt{m_n^*+p_\perp^2}-s\mu_Nk_pB)^2}+g_\omega\omega^0-\frac{1}{2}g_\rho\rho^0, \end{aligned} \tag{4}$$

where $\nu = n+\frac{1}{2}-\text{sgn}(q_b)\frac{s}{2} = 0, 1, 2, ...$ are the Landau levels for the fermions with electric charge q, s is the quantum number of spin e assume values $+1$ for spin up e -1 for spin down cases.

Defining $\overline{m}_n = m^* - s\mu_N k_n B$, the Fermi momenta $p^p_{F,\nu,s}$ of the protons and $p^n_{F,s}$ of the neutrons and their relationship with the Fermi energies of protons E^p_F and neutrons E^n_F can be written as:

$$p^p_{F,\nu,s} = E^{p2}_F - [\sqrt{m^{*2}_p + 2\nu q_p B} - s\mu_N k_p B]^2, \qquad p^n_{F,s} = E^n_F - \overline{m}^2_n. \tag{5}$$

The summation over the Landau level ν is made until ν_{max}, which is the largest value of ν for which the square of Fermi momenta of the particle is still positive and corresponds to the closest integer, from below to:

$$\nu_{max} = \left[\frac{(E^p_p + s\mu_N k_p B)^2 - m^*_p}{2|q_p|B}\right] \tag{6}$$

The chemical potentials of protons and neutrons are:

$$\mu_p = E^p_F + g_\omega \omega^0 + \frac{1}{2} g_\rho \rho^0, \qquad \mu_n = E^n_F + g_\omega \omega^0 - \frac{1}{2} g_\rho \rho^0. \tag{7}$$

The total energy density of the neutron star matter is given by:

$$\begin{aligned} \varepsilon_t = & \sum_{b=p,n} \varepsilon_b + \frac{1}{2} m_\sigma \sigma^2_0 + \frac{1}{2} m_\omega \omega^2_0 + \frac{1}{2} m_\rho \rho^2_0 \\ & + \frac{1}{8} \xi (g_\omega \omega_0)^4 + \frac{1}{3} b m_n (g_\sigma \sigma_0)^3 + \frac{1}{4!} c (g_\sigma \sigma_0)^4 + 3\Lambda_\omega (g_\omega \omega_0)^2 (g_\rho \rho_0)^2, \end{aligned} \tag{8}$$

where the expressions for the proton and neutron energy density are, respectively,

$$\begin{aligned} \varepsilon_p = & \sum_{n=0}^{n_{max}} \sum_s \frac{|Q_p|B}{4\pi^2} \Bigg[p^p_{F,n,s} E^p_F + (\sqrt{m^2_p + 2Q_p B n} - s\mu_N k_p B)^2 \\ & \times ln \Bigg| \frac{p^p_{F,n,s} + E^p_F}{(\sqrt{m^2_p + 2Q_p B n} - s\mu_N k_p B)} \Bigg| \Bigg], \\ \varepsilon_n = & \frac{1}{4\pi^2} \sum_s \frac{1}{2} E^{n3}_F p^n_{F,n,s} - \frac{2}{3} s\mu_N k_n B E^{n3}_F \left(arc\sin \frac{\overline{m}}{E^n_F} - \frac{\pi}{2} \right) \\ & - \left(\frac{1}{3} s\mu_N k_n B + \frac{1}{4} \overline{m} \right) \left[\overline{m} p^n_{F,n,s} E^n_F + \overline{m}^3 ln \left(\left| \frac{E^n_F + p^n_{F,n,s}}{\overline{m}} \right| \right) \right]. \end{aligned} \tag{9}$$

and the pressure for the baryons is $P_b = n_b \mu_b - \varepsilon_b$ (b=p,n).

The symmetry energy and its density slope are, respectively, defined as:

$$a_{sym} = \frac{1}{2} \frac{\partial^2 E/A}{\partial t^2}, \qquad L = 3n_0 \frac{\partial a_{sym}}{\partial n}, \tag{10}$$

where $n_0 = 0.153\ fm^{-3}$ is the saturation density of nuclear matter and $t = \frac{n_n - n_p}{n_p + n_n}$ is the asymmetry parameter.

3. Results and discussions

In figure (1) we show how the behaviour of the symmetry energy, given in (10), varies with the baryonic density when different intensities of magnetic fields are applied. $B = 0\ G$ is used as a field of reference.

On the left side, the upper and lower panels represent the NL3 and FSU models for $B = 10^{17}\ G$. The effect of Landau levels is seen, but due to the relatively low magnetic field, the curve of the symmetry energy is very similar to that of the $B = 0\ G$ field. The irregularities are due to the filling of Landau levels, which have small energy differences between them for fields of this intensities.

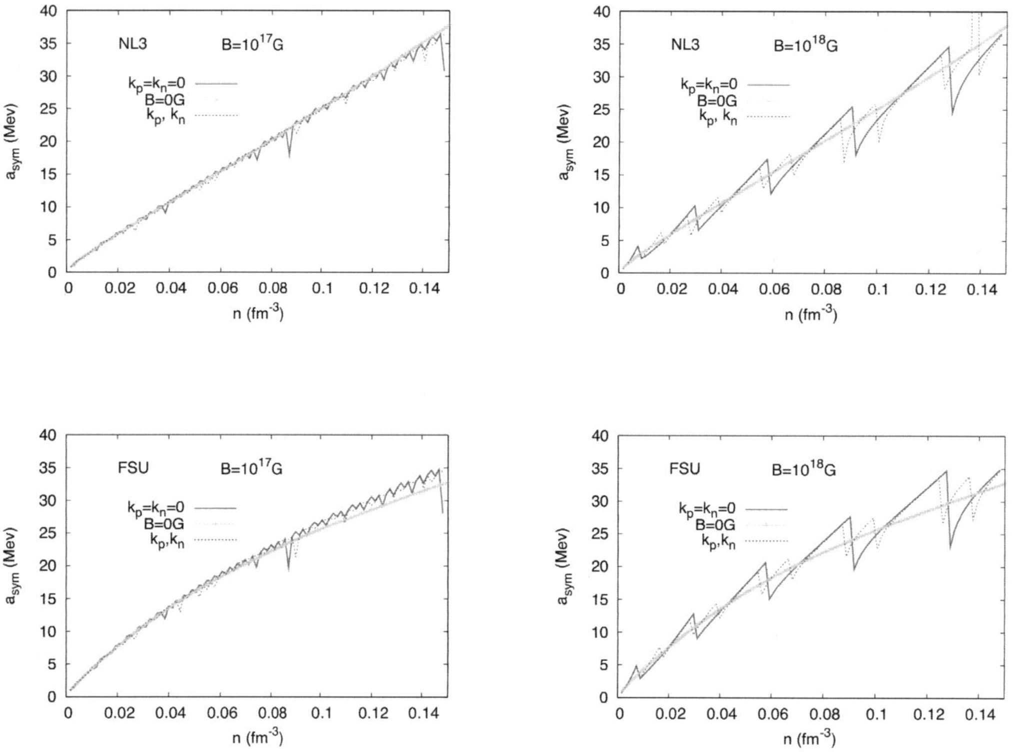

Figure 1. Symmetry energy versus baryonic density for NL3 and FSU models, with and without anomalous magnetic moment. The left panels presents $B = 0$ and $10^{17}\ G$ and the right panels presents $B = 0$ and $10^{18}\ G$ values for the magnetic field.

On the right side the curves for $B = 10^{18}\ G$ field are shown. As seen from equation (6) a smaller number of Landau levels is filled, due to the higher value of magnetic field. The jumps obtained are larger than those arising with $B = 10^{17}\ G$. This behaviour will affect the chemical composition of stellar matter. In particular, we get regions where the symmetry energy becomes larger implying larger proton fractions in β-equilibrium matter. The opposite occurs in neighbouring regions. The magnitude $B = 10^{18}\ G$ may be larger than expected at the inner crust of a compact star, but it would be interesting to investigate the effect of B on the pasta phase. Since the effects depend on the filling of the Landau levels and, therefore, on the density, the different pasta phase geometries may be affected in different ways.

Journal of Physics: Conference Series **342** (2012) 012002 doi:10.1088/1742-6596/342/1/012002

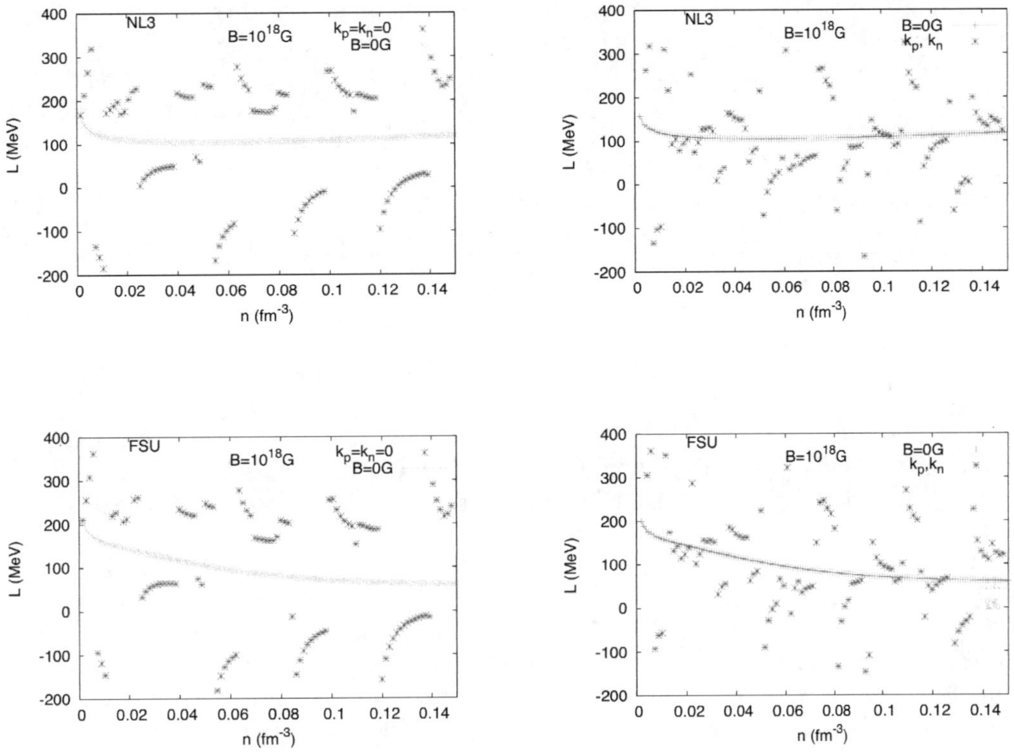

Figure 2. Density slope versus baryonic density for NL3 and FSU models, for magnetic fields of magnitude $B = 0, 10^{18}\ G$. In the left upper and lower panels, anomalous magnetic moment are not considered. In the right they are included.

Comparing results for FSU and NL3 there are different trends due to different behavior of the symmetry energy with density in bot models. For NL3, the symmetry energy for $B = 10^{17}\ G$ has a more linear dependence on the density which follows the $B = 0\ G$ line tendency. This also happens for the $B = 10^{18}\ G$ scenario, but because of the smaller number of Landau levels the symmetry energy curve presents larger oscillations around de $B = 0\ G$ line. The FSU model shows a similar behavior, both for the $B = 10^{17}\ G$ and $B = 10^{18}\ G$ curves, but the first one demonstrate a small discrepancy around the saturation density.

The inclusion of the anomalous magnetic moment has almost no effect for $B = 10^{17}\ G$. However, for a larger field such as $B = 10^{18}\ G$ a clear effect is observed due to the spin polarization of neutrons and protons, on top of the Landau level quantization effect on protons. The main result is an increase of the number and a decrease of the magnitude of the discontinuities around the average $B = 0$ result.

The slope of the symmetry energy is defining how strong is the distillation effect and the magnitude of the neutron drip in the inner crust pasta phase. Figure (1) has shown that the slope of the symmetry energy is also affected by the magnetic field and, therefore, we will next calculate and discuss this quantity.

In figure (2) the slope of the symmetry energy, eq. (10), is presented for both models, without (left panel) and with (right panel) the inclusion of the anomalous magnetic moment. We again use the $B = 0\ G$ as reference. As expected, the Landau levels generated by the magnetic fields have a strong effect on the slope, with regions where it is clearly larger, followed by regions where it is smaller, becoming even negative. In the regions where the slope of the symmetry energy becomes larger the isospin distillation effect will be stronger, and as a result, more isospin symmetric paste clusters are obtained. A second effect will be a denser neutron background gas. The opposite will occur in the neighbouring regions with a smaller slope L. These results show that the properties of the clusters will be affected by the magnetic field, and, as a result all transport properties that depend on the structure of this phase.

The inclusion of the anomalous magnetic moment will bring extra complexity to the problem with a less systematic behaviour due to the spin polarization effects.

4. Summary

In the present work, we have studied the behavior of the symmetry energy and its slope. In particular, we have compared these quantities obtained for two different models, NL3 and FSU, for three values of magnetic field, $B = 0$, 10^{17}, $10^{18}\ G$ and under two different conditions, with and without the inclusion of the anomalous magnetic moment.

It was shown that the appearance of Landau levels due to the magnetic field will affect the symmetry energy, giving rise to discontinuities: the symmetry energy may become larger or smaller than the $B = 0$ reference for given densities which are determined with the filling of Landau levels or spin levels. It is expected that the observed behaviour will affect the structure of the inner crust of a neutron star, namely the pasta phase. Not only the fraction of protons in β-equilibrium matter will change but also the properties of the clusters and the background neutron gas. The effects are larger the more intense the magnetic field. The inclusion of the anomalous magnetic moment will only be noticed for quite high fields and will make the behaviour even more complex.

5. Acknowledgments

This work was supported by CAPES/FCT 232/09.

References

[1] Ban S., Meng J., Satula W., Wyss R., Phys. Lett. B 633, 231-236 (2006).
[2] Horowitz C.J., Piekarewicz J., Phys. Rev. C 64, 062802 (2001)
[3] Chang Xu, Bao-An Li, Lie-Wen Chen, Phys. Rev. C 82, 054607 (2010)
[4] Rabhi A., Providência C., and Da Providência J., J. Phys. G: Nucl. Part. Phys. 37 075102 (2010).
[5] Broderick A., Prakash M., Lattimer M., ApJ 537, 351-367 (2000).
[6] Fattoyev F. J., Horowitz C. J., Piekarewicz J., and Shen G., Phys. Rev. C 82, 055803 (2010).
[7] Horowitz C.J., Piekarewicz J., Phys. Rev. Lett. 86, 25 (2001)
[8] Rabhi A., Providência C., and Da Providência J., J. Phys. G: Nucl. Part. Phys. 35 125201 (2008).

Unified equation of state for neutron stars and supernova cores using the nuclear energy-density functional theory

A. F. Fantina[1], N. Chamel[1], J. M. Pearson[2] and S. Goriely[1]
[1]Insitut d'Astronomie et d'Astrophysique, Université Libre de Bruxelles, CP226, B-1050 Brussels, Belgium
[2]Département de Physique, Université de Montréal, Montréal, Québec, Canada H3C 3J7
E-mail: afantina@ulb.ac.be

Abstract. We present a unified approach to the equation of state (EoS) of dense matter at any temperature, based on the nuclear energy-density functional (EDF) theory. Both homogeneous and inhomogeneous phases can be treated consistently. In particular, we have constructed three different EoSs of cold catalyzed matter for a wide range of densities from $\sim 10^5$ g cm^{-3} to $\sim 10^{15}$ g cm^{-3}. For this purpose, we have employed generalized Skyrme functionals fitted to essentially all experimental nuclear mass data and constrained to reproduce properties of homogeneous nuclear matter as obtained from many-body calculations. We have applied these unified EoSs to compute the structure of cold isolated neutron stars (NSs).

1. Introduction

Type II Supernovae (SNe) represent the end point of stellar evolution for stars with a mass greater than about 8 $M_\odot$. When the core of massive stars exceeds the Chandrasekhar mass, it undergoes a gravitational collapse followed by a bounce leading to the formation of a shock wave which expels the outer layers of the star and triggers the explosion. The core shrinks and its density increases up to a few times the nuclear matter saturation density, thus producing a dense compact remnant: a NS or a black hole. Because of the wide range of conditions encountered in the collapsing core (densities from 10 to 10^{15} g cm^{-3}, temperatures from 0 to 100 MeV, and electron fraction Y_e from 0 - pure neutron matter - to 0.6), a unified description of the matter is a very challenging task. Indeed, the interiors of SN cores and NSs are expected to exhibit very different phases of matter, from ordinary nuclei to homogeneous nuclear matter. The EDF theory allows for a consistent and computationally tractable treatment of these various phases, both at zero and at finite temperatures. At very high densities, matter might contain other particles like hyperons or even deconfined quarks (see e.g. Ref. [1]). The EDF theory can be extended to take into account these additional particles, but here we will only consider nucleons and leptons. We have constructed a set of three different unified EoSs describing all regions of the NS, from the surface to the core, as described in the next section.

2. Construction of the energy-density functional

The EDFs that we consider here are based on generalized Skyrme effective nucleon-nucleon interactions [2, 3]:

$$\begin{aligned} v_{ij} &= t_0(1+x_0P_\sigma)\delta(\mathbf{r}_{ij}) + \frac{1}{2}t_1(1+x_1P_\sigma)\frac{1}{\hbar^2}\left[p_{ij}^2\,\delta(\mathbf{r}_{ij}) + \delta(\mathbf{r}_{ij})\,p_{ij}^2\right] \\ &+ t_2(1+x_2P_\sigma)\frac{1}{\hbar^2}\mathbf{p}_{ij}.\delta(\mathbf{r}_{ij})\,\mathbf{p}_{ij} + \frac{1}{6}t_3(1+x_3P_\sigma)\,n(\mathbf{r})^\alpha\,\delta(\mathbf{r}_{ij}) \\ &+ \frac{1}{2}t_4(1+x_4P_\sigma)\frac{1}{\hbar^2}\left[p_{ij}^2\,n(\mathbf{r})^\beta\,\delta(\mathbf{r}_{ij}) + \delta(\mathbf{r}_{ij})\,n(\mathbf{r})^\beta\,p_{ij}^2\right] \\ &+ t_5(1+x_5P_\sigma)\frac{1}{\hbar^2}\mathbf{p}_{ij}.n(\mathbf{r})^\gamma\,\delta(\mathbf{r}_{ij})\,\mathbf{p}_{ij} + \frac{\mathrm{i}}{\hbar^2}W_0(\boldsymbol{\sigma_i}+\boldsymbol{\sigma_j})\cdot\mathbf{p}_{ij}\times\delta(\mathbf{r}_{ij})\,\mathbf{p}_{ij} \quad , \end{aligned} \tag{1}$$

supplemented with a microscopic contact pairing interaction [3, 4]. In Eq. (1), $\mathbf{r}_{ij} = \mathbf{r}_i - \mathbf{r}_j$, $\mathbf{r} = (\mathbf{r}_i + \mathbf{r}_j)/2$, $\mathbf{p}_{ij} = -\mathrm{i}\hbar(\nabla_i - \nabla_j)/2$ is the relative momentum, P_σ is the two-body spin-exchange operator, and $n(\mathbf{r}) = n_n(\mathbf{r}) + n_p(\mathbf{r})$ is the total local density, $n_n(\mathbf{r})$ and $n_p(\mathbf{r})$ being the neutron and proton densities, respectively. The t_4 and t_5 terms are the density-dependent generalizations of the t_1 and t_2 terms, respectively. These EDFs were fitted to the 2149 measured masses of nuclei with N and $Z \geq 8$ given in the 2003 AME [5], with an rms deviation of 0.58 MeV. These EDFs are therefore well-suited for describing the properties of neutron-rich nuclei found in the outer crust of a NS. The masses were obtained by adding to the Hartree-Fock-Bogoliubov (HFB) energy a phenomenological Wigner term and correction term for the spurious collective energy. The EDFs were simultaneously constrained to reproduce various properties of homogeneous nuclear matter as obtained from many-body calculations using realistic two- and three- nucleon interactions. Three different functionals BSk19, BSk20 and BSk21 have been constructed, constrained to reproduce three different neutron matter EoSs at $T = 0$ thus reflecting the current lack of knowledge of the high-density behaviour of nuclear matter (see Fig. 1 in Ref. [3]). In particular, BSk19 was adjusted to the softest EoS of neutron matter known to us, the EoS of Friedman and Pandharipande obtained from the realistic Urbana v_{14} nucleon-nucleon force with the three-body force TNI [6], BSk20 was fitted to the EoS of Akmal *et al.* labeled "A18 + δv + UIX" in Ref. [7], whereas BSk21 was constrained to reproduce the very stiff "V18" EoS from Li and Schulze [8]. Furthermore, i) spurious spin and spin-isospin instabilities in nuclear matter that generally plague existing Skyrme functionals have been eliminated for all densities and temperatures prevailing in NS and SN cores [9], ii) a qualitatively realistic distribution of the potential energy among the four spin-isospin channels in nuclear matter has been obtained, iii) the isovector effective mass is found to be smaller than the isoscalar effective mass, in agreement with both experiments and many-body calculations. With these features, our EDFs can be reliably applied to all regions of the NS: the outer crust, the inner crust, and the liquid core.

In Fig. 1, we have plotted for our three functionals the free energy per baryon and the pressure of hot neutron matter, as a function of the baryon number density. For comparison, the many-body calculations at $T = 0$ from Friedman and Pandharipande [6], Akmal *et al.* [7], Li and Schulze [8] discussed previously, together with microscopic calculations at finite temperature based on the Self-Consistent Green Function (SCGF) method with CD Bonn potential [10] are also shown. As can be seen in Fig. 1, our functionals are in close agreement with these calculations, at least for densities up to nuclear saturation density, even though they were only fitted to realistic EoSs of neutron matter at zero temperature. We therefore conclude that our functionals are also suitable for describing the hot dense matter found in SN cores and proto-neutron stars.

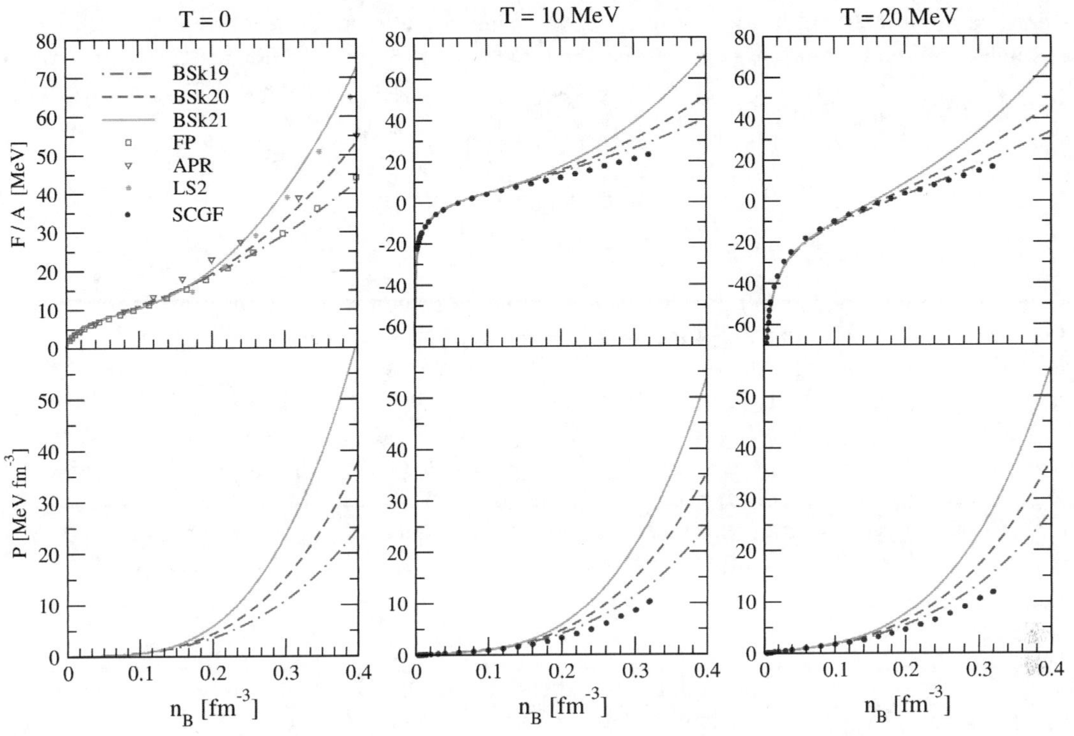

Figure 1. Free energy per baryon (upper panels) and pressure (lower panels) of neutron matter as a function of the number baryon density, at temperatures $T = 0$, 10, 20 MeV. The microscopic calculations at $T = 0$ are taken from Friedman and Pandharipande [6] (labeled as "FP"), Akmal *et al.* [7] (labeled as "APR"), and Li and Schulze [8] (labeled as "LS2"), while those based on Self-Consistent Green Function with CD Bonn potential (labeled as "SCGF") are taken from Ref. [10].

3. Unified equation of state of neutron stars

3.1. The microscopic model

We recognize in the NS three distinct regions: i) the outer crust where nuclei are arranged in a body-centered lattice and coexist with a gas of relativistic electrons, ii) the inner crust where nuclei coexist with both electrons and unbound neutrons and iii) the liquid core which consists of a uniform mixture of nucleons and leptons. All regions were treated consistently using the same EDF at $T = 0$ under the assumption of β equilibrium. The EoS of the outer crust was calculated in the framework of the BPS model [11] using the latest experimental atomic masses complemented with theoretical masses obtained coherently from our HFB mass model (see Ref. [12] for details). With increasing density, the equilibrium nuclei become more and more neutron-rich until neutrons start to drip out. This transition, which defines the boundary between the outer crust and the inner crust, occurs when the neutron chemical potential exceeds the neutron rest mass. For the inner crust, we employed the temperature-dependent extended Thomas-Fermi method (up to the 4th order) with proton shell effects added via the Strutinsky integral theorem [13]. This method, originally developed for calculating the EoS of SN cores, is a computationally fast approximation to the full Hartree-Fock equations. Neutron shell corrections, which are known to be much smaller than proton shell corrections [14], are neglected. In order to further reduce the computational work, we assumed that nuclear clusters are spherical and we made use of the Wigner-Seitz approximation. In addition, the local nucleon number densities were parametrized as ($q = n$ or p for neutrons and protons respectively) $n_q(r) = n_{Bq} + n_{\Lambda q} f_q(r)$, in which n_{Bq} is a constant background term, while $f_q(r)$ has the form of a modified Fermi distribution [15] given by

$$f_q(r) = \frac{1}{1 + \exp\left\{\left(\frac{C_q - R}{r - R}\right)^2 - 1\right\} \exp\left(\frac{r - C_q}{a_q}\right)} \quad , \tag{2}$$

R being the radius of the Wigner-Seitz cell. The parameters C_q determine the size of the clusters whereas a_q take into account the diffuseness of the clusters surface. In this "damped" form of the usual simple Fermi profile all density derivatives vanish at the surface of the cell, thereby ensuring a smooth matching of the nucleon distributions between adjacent cells, and also satisfying the necessary conditions for the validity of the underlying semi-classical expressions. The EoS was obtained by minimizing, at a fixed average baryon density $\bar{n}$, the free energy per nucleon f. This procedure is equivalent to the minimization of the Gibbs free energy per nucleon g but the former is computationally much more convenient. At zero temperature, f coincides with the internal energy per nucleon e which can be expressed as (see Refs. [13, 15] for details)

$$e = e_{\rm nuc} + e_e + e_C - Y_e\, Q_{n,\beta} \quad , \tag{3}$$

where e_e is the electron kinetic energy per nucleon, e_C the Coulomb energy per nucleon and $Y_e Q_{n,\beta}$ takes into account the neutron-proton mass difference ($Q_{n,\beta} = 0.728$ MeV is the neutron β-decay energy and Y_e the electron fraction). The nuclear energy $e_{\rm nuc}$ is given by

$$e_{\rm nuc} = \frac{4\pi}{A} \int_{\rm cell} r^2 \mathcal{E}_{\rm Sky}^{\rm ETF}(r) dr + e_{\rm sh}^p \quad , \tag{4}$$

where A is the total number of nucleons in the Wigner-Seitz cell, $\mathcal{E}_{\rm Sky}^{\rm ETF}(r)$ is the ETF approximation to the energy density $\mathcal{E}_{\rm Sky}(r)$ given by the generalized Skyrme functional. Proton shell effects were added perturbatively using the Strutinsky integral method and are contained in $e_{\rm sh}^p$. The minimization of e with respect to the eight parameters (Z, N, plus the three parameters of the Wigner-Seitz cell for each nucleon type) was performed using the CERN routine MINUIT.

The EoS of the core was calculated under the assumption of a uniform plasma of neutrons, protons, electrons and muons in β equilibrium.

3.2. Results

In Fig. 2, we show the EoSs corresponding to the three different functionals BSk19, BSk20 and BSk21 for all regions of a NS. For practical purpose, we have fitted our EoSs using the analytical representation proposed in Eq. (14) of Ref. [16]. The new coefficients of the fit are calculated using a least-squares method. The fits reproduce the tabulated EoSs with an error below 4% except near the crust-core boundary where the error may reach about 10%.

We have applied our unified EoSs to compute the mass-radius relationship of non-rotating NSs. The NS structure is determined using the LORENE library [17, 18]. Figure 3 shows the gravitational mass as a function of the circumferential radius of static NSs for our three EoSs. In particular, the maximum NS masses for our three different functionals BSk19, BSk20 and BSk21 are 1.86 $M_\odot$, 2.15 $M_\odot$, 2.28 $M_\odot$ respectively. The shaded area corresponds to the largest (up-to-date) mass ($1.97 \pm 0.04\ M_\odot$) of the recently observed millisecond pulsar J1614−2230 [19]. This observation rules out the softest EoS BSk19.

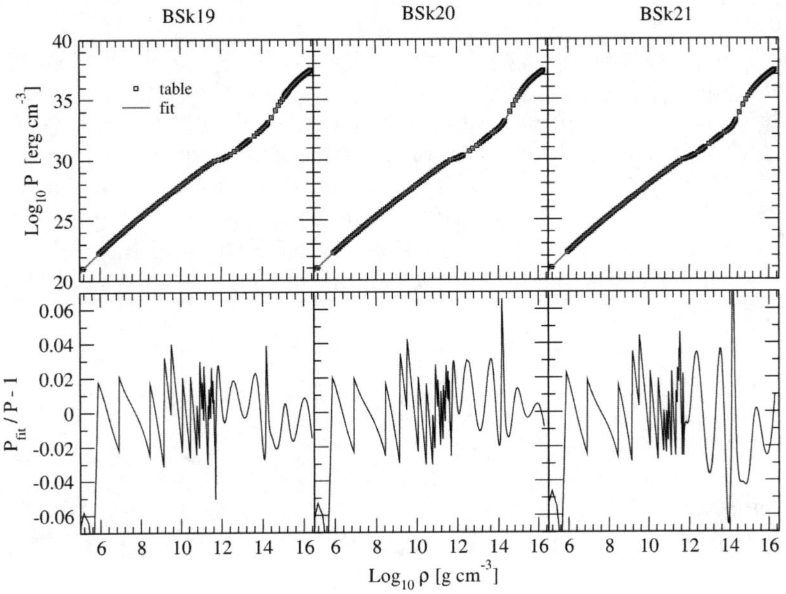

Figure 2. Upper panel: pressure P versus mass-energy density ρ of cold catalyzed matter. Rarefied tabular data are indicated by symbols, while the solid lines show the analytical representation for non-rotational configurations discussed in the text. Lower panel: relative difference between the data and the fit.

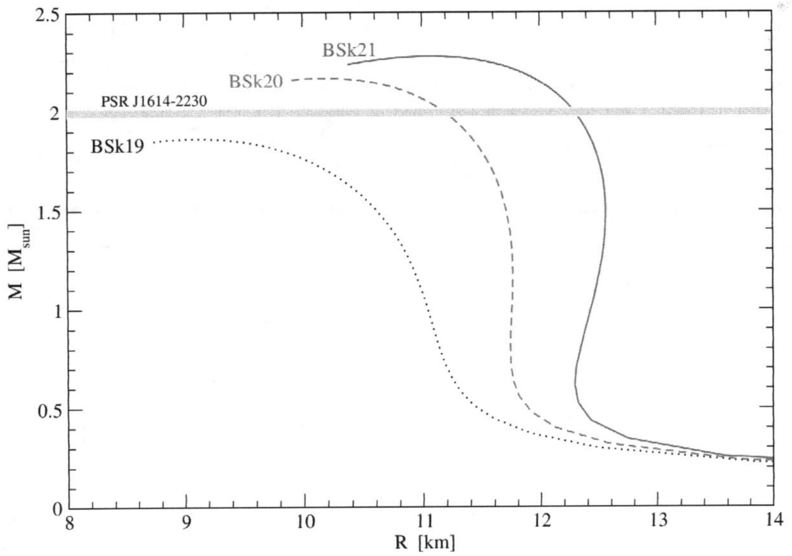

Figure 3. Gravitational mass versus circumferential radius for non-rotating neutron stars. The configurations are calculated using three different equations of state, namely BSk19, BSk20 and BSk21.

4. Conclusion

We have presented three different unified EoSs of NS based on the EDF theory. These functionals are based on generalized Skyrme forces which were fitted to essentially all experimental nuclear mass data and, in addition, were constrained to reproduce several properties of homogeneous nuclear matter as obtained from microscopic calculations using realistic nucleon-nucleon potentials. These EoSs have also been applied to compute the static structure of NSs. We have found that the EoSs BSk20 and BSk21 are compatible with the recent observation of a high-mass NS. Although we have presented the EoSs for the limiting case of cold catalyzed matter as relevant for cold non-accreting NS, our method can be easily extended to the hot dense matter of SN cores.

Acknowledgments

A.F.F. would like to acknowledge J. Novak for very helpful discussions. This work has been supported by FNRS (Belgium), the Communauté française de Belgique (Actions de Recherche Concertées), NSERC (Canada) and CompStar, a Research Networking Programme of the European Science Foundation.

References

[1] Haensel P., Potekhin A. Y., and Yakovlev D. G. 2007, *"Neutron Stars I. Equation of State and Structure", Astrophysics and Space Science Library*, **326**
[2] Chamel N., Goriely S., and Pearson J. M. 2009, *Phys. Rev. C*, **80**, 065804
[3] Goriely S., Chamel N., and Pearson J. M. 2010, *Phys. Rev. C* **82**, 035804
[4] Chamel N. 2010, *Phys. Rev. C* **82**, 014313
[5] Audi G., Wapstra A. H., and Thibault C. 2003, *Nucl. Phys.* **A 729**, 337
[6] Friedman B., and Pandharipande V. R. 1981, *Nucl. Phys.* **A 361**, 502
[7] Akmal A., Pandharipande V. R., and Ravenhall D. G. 1998, *Phys. Rev. C* **58**, 1804
[8] Li Z. H., and Schulze H. J. 2008, *Phys. Rev. C* **78**, 028801
[9] Chamel N., and Goriely S. 2010, *Phys. Rev. C* **82**, 045804
[10] Rios A., Polls A., and Vidaña I. 2009, *Phys. Rev. C* **79**, 025802
[11] Baym G., Pethick C., and Sutherland P. 1971, *Astrophys. J.*, **170**, 299
[12] Pearson J. M., Goriely S., and Chamel N. 2011, *Phys. Rev. C* **83**, 065810
[13] Onsi M., Dutta A. K., Chatri H., Goriely S., Chamel N., and Pearson J. M. 2008, *Phys. Rev. C* **77**, 065805
[14] Chamel N., Naimi S., Khan E., and Margueron J. 2007, *Phys. Rev. C* **75**, 055806
[15] Pearson J. M., Chamel N., Goriely S. and Ducoin C., submitted to *Phys. Rev. C*
[16] Haensel P., and Potekhin A. Y. 2004, *Astron. Astrophys.* **428**, 191
[17] http://www.lorene.obspm.fr
[18] Gourgoulhon E. 2010, *Preprint* arXiv:1003.5015
[19] Demorest P. B., Pennucci T., Ransom S. M., Roberts M. S. E., and Hessels J. W. T. 2010, *Nature* **467**, 1081

Vortex-lattice interaction in Pulsar Glitches

Fabrizio Grill[1,2] and Pierre Pizzochero[2,3]

[1] Centro de Física Computacional, Department of Physics, University of Coimbra, PT-3004-516 Coimbra, Portugal
[2] Dipartimento di Fisica, Universitá degli Studi di Milano, Via Celoria 16, 20133 Milan, Italy
[3] Istituto Nazionale di Fisica Nucleare, sezione di Milano, Via Celoria 16, 20133 Milano, Italy

E-mail: fabrizio.grill@gmail.com

Abstract. In this proceeding we present a quantitative model to study the pinning interaction between a neutron vortex and the nuclear lattice in the inner crust of a neutron star, which is formed by nuclear clusters immersed in a superfluid neutron gas. This is one of the most important quantities in the vortex model for pulsar glitches, where the sudden variations of the rotation velocity observed in some neutron stars are explained in terms of pinning-depinning processes between vortices and the nuclear lattice. The study has been done through an actual counting of the number of pinning sites between the vortex and the nuclear lattice. Moreover, it has been recognized that the vortices can have a generic orientation with respect to the lattice and thus the simple pinned-unpinned scenario (that is characteristic of the aligned case and has been assumed in previous qualitative models) is not acceptable for a realistic calculation. The forces that we have found are more than two orders of magnitude smaller than those calculated so far in the literature, and of the same order as what suggested to explain real Vela-like glitches. They can thus be applied to quantitative models for the vortex dynamics, in order to better simulate and understand these remarkable astrophysical phenomena.

1. Introduction

The rotation of a pulsar is characterized by a regular slow down, which is due to the emission of electromagnetic and gravitational waves. However, in the rotation of many pulsars sudden increases in the spin rate, known as glitches, have been observed. These peculiar phenomena are very important in the framework of the study of nuclear matter, because from their observation it may be possible to extract informations about some matter properties. Indeed they have been observed also in isolated stars and so they can only be related to the behavior of matter in the neutron star interior.

The origin of these phenomena is still debated, but if the smaller glitches observed in younger and hotter stars (e.g. Crab) appear to be linked to starquakes [14], the giant glitches observed in older and colder pulsars (e.g. Vela) seem to be connected to the presence of bulk superfluidity inside these stars. Indeed, the neutron superfluid cannot follow the rigid rotation of the normal (i.e. non-superfluid) component of the pulsar, since the superfluid flow must be irrotational ($\nabla \times \vec{v}_s = 0$). Instead, it develops an array of vortices, each carrying angular momentum, with macroscopic average velocity field satisfying the rigid rotation relation ($\nabla \times \langle \vec{v}_s \rangle = 2\vec{\Omega}_s$) and macroscopic angular velocity proportional to the vortex density ($\Omega_s \propto n_{\rm v}$).

According to the 'vortex model' of Anderson and Itoh [2], the glitch phenomenon can be the natural macroscopic outcome of the interaction between these vortices and the Coulomb

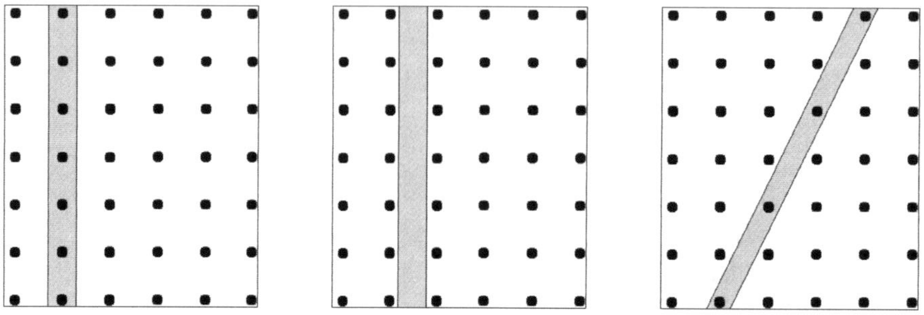

Figure 1. Representation of some possible vortex-lattice configurations. a and b are the aligned configurations, respectively vortex bound and free, while c is a possible non-aligned configuration.

lattice of neutron-rich nuclear clusters, which is predicted to coexist with the neutron superfluid in the inner crust [15]. This pinning interaction is expected to bind vortices to the normal component of the star, thus freezing the vortex distribution and then fixing the superfluid angular velocity. As the normal matter slows down, the difference between the angular velocity of the two components of the pulsar grows continuously. This differential motion will be the source of a lift force, called the Magnus force, whose intensity per unit length is proportional to the angular velocity difference ($f_{mag} \propto \Omega_s - \Omega_n$). Only when the total Magnus force equals the total pinning force, the vortex is detached from the lattice (depinning). Thus free to move, it can transfer its angular momentum to the normal component of the star under the action of drag forces. According to the vortex model, giant glitches are due to the simultaneous depinning of a large number of vortices, which is followed by the rapid transfer of their angular momentum. Furthermore such a storage and trigger mechanism would have a natural periodicity, as indeed observed in Vela [5].

In order to make a quantitative study of the pulsar glitch phenomenon, it is crucial to perform an accurate comparison between the total pinning and Magnus forces acting on the vortex line. In this framework, several models have been proposed in order to calculate the pinning interaction between the vortex and a single nucleus [3, 4, 6, 7, 8, 13]; on the other hand, the Magnus force is a classical hydrodynamical force and so it is well known since the nineteenth century. Anyways, a direct comparison between pinning and Magnus forces is still impossible because the Magnus force is given per unit length, while in the cited works the pinning force is calculated per nuclear site: thus an accurate calculation of the pinning force per unit length is still necessary.

In the literature, the comparison between pinning and Magnus forces has been done several times, but using a simple qualitative calculation of the number of pinning sites per unit length (e.g. [1]). Moreover, these calculations were based on the assumption that the vortices are aligned with the lattice, thus considering just two possible configurations, the vortex totally bound (Fig. 1a) and the vortex totally free (Fig. 1b). In this very simple scenario, the pinning force would be proportional to the number of the pinning sites of the bound configuration ($F \propto N$). This assumption has the merit to make the calculations simpler and faster, but the resulting pinning forces are two or three orders of magnitude larger than those suggested to explain real Vela-like glitches [16].

On the other hand, in the nineties Jones [10, 11, 12] proposed a study of the number of pinning sites per unit length with a generic orientation between the vortex and the lattice (Fig. 1c). In this condition, there are not necessarily configurations in which the vortex is totally free, i.e. no pinning sites with the lattice, and the pinning force can be calculated as proportional to the difference between the maximum and the minimum number of pinning sites

Journal of Physics: Conference Series **342** (2012) 012004 doi:10.1088/1742-6596/342/1/012004

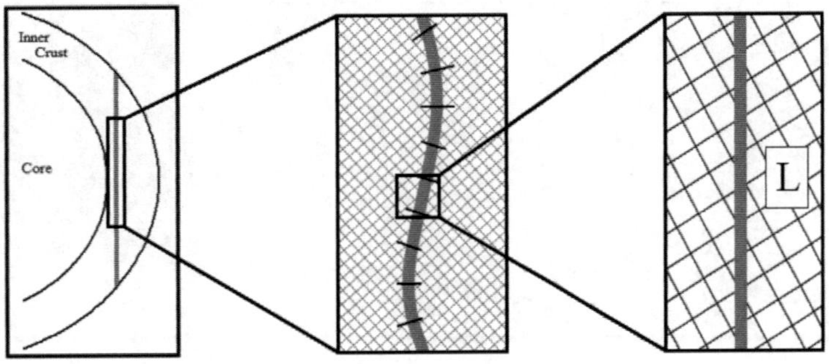

Figure 2. Representation of the vortex partition.

($F \propto N_{max} - N_{min}$). With this model Jones found that the intensity of the pinning force per unit length would decrease as the considered vortex length increases. He thus concluded that, if the rigid portion of vortex is long enough, the pinning force per unit length becomes negligible.

In this proceeding we present the first quantitative estimate of the pinning force per unit length. In the section 2 we will describe the calculation, while in the section 3 we will present the results obtained with our model.

2. The pinning force per unit length

The first quantity necessary for the pinning force calculation is the vortex length on which perform the study. Indeed, if the estimate done by Jones turns out to be correct, the vortex model would no longer explain the glitch behavior. However, in this and all other calculations found in the literature, the vortex has always been considered as unbendable. But its rigidity is not infinite and so in the star it will bend: thus, it is not correct to perform a calculation on the whole vortex, but it is necessary to develop the study dividing it in smaller parts. The length of each part is calculated to be the maximum length so that it can still be considered as rigid (Fig. 2). Its value is obtained by assuming that the vortex, under tension T, will bend under the influence of the pinning force, and thus equating the energy of two limiting configurations: the rigid vortex and the vortex that has bent in order to have an additional pinning site. With this very simple calculation, we have found that the maximum vortex length is approximately $L \approx 1000\ R_{ws}$, where R_{ws} is the radius of the Wigner-Seitz cell.

In this condition, the pinning force that acts on the entire vortex can be found by summing the contribution of each rigid portion

$$F = \sum \langle F_L \rangle \tag{1}$$

where the rigid contributions have been calculated as the mean over all the possible orientations between the vortex and the lattice, orientation defined by standard spherical coordinates (θ, φ) with solid angle $d\Omega$:

$$\langle F_L \rangle = \frac{\int F_L(\theta, \varphi) d\Omega}{\int d\Omega} \tag{2}$$

Finally, in order to have a more useful way to calculate the pinning force, we rewrite the expression (1) as an integral relation

$$F = \sum \langle F_L \rangle = \int \langle F \rangle \, dL \tag{3}$$

where $\langle F \rangle = \frac{\langle F_L \rangle}{L}$ is the pinning force per unit length.

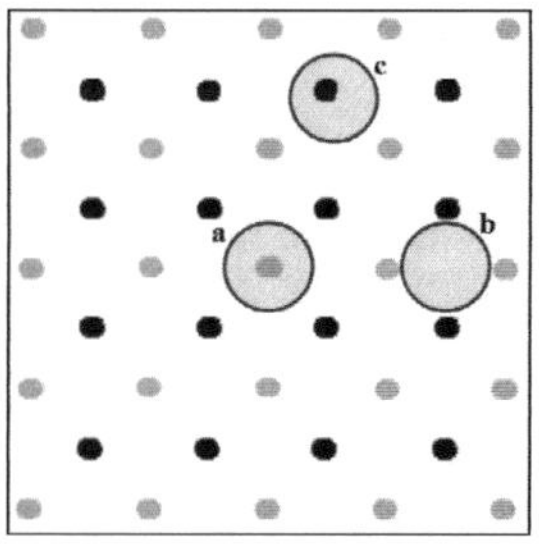

Figure 3. Representation of the translations for an aligned vortex. We show the initial condition (a) and two possible translations perpendicular to the vortex line (b and c).

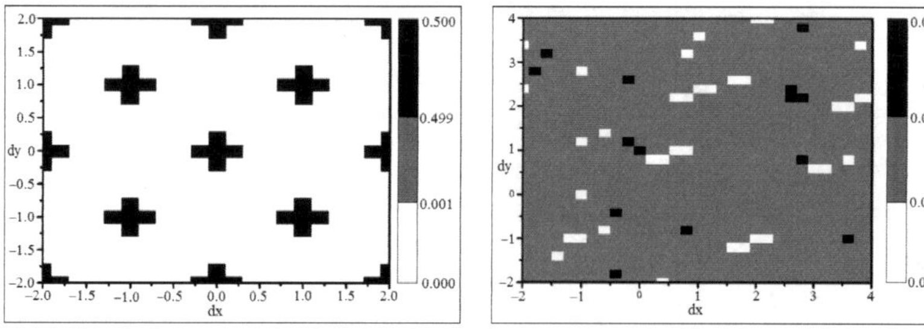

Figure 4. Number of pinning sites in unit of the Wigner-Seitz radius. We show two possible cases: the vortex aligned (left) and the vortex non-aligned (right) with the lattice. The number of pinning sites is describe in the sidebar. dx and dy represent the translation of the vortex with respect to the initial condition. It is important to notice that, as Jones supposed, the minimum number of pinning sites in the non-aligned configuration is not zero.

In order to use equation (2) for the calculation of the pinning force per unit length, it is necessary to evaluate the quantity $F_L(\theta,\varphi)$. Its calculation is done through a classical approach

$$F_L(\theta,\varphi) \propto \frac{\Delta E(\theta,\varphi)}{D(\theta,\varphi)} = \frac{E_p\left(N_{free}(\theta,\varphi) - N_{bound}(\theta,\varphi)\right)}{D(\theta,\varphi)} \tag{4}$$

where $\Delta E(\theta,\varphi)$ is the energy difference between the configurations on which the vortex is free and bound, which are respectively characterized by a number of pinning sites equal to N_{free} and N_{bound}, E_p is the pinning energy per site and $D(\theta,\varphi)$ is the distance between the vortex position of these two configurations.

The study of the number of pinning sites in the two configurations, bound and free, is done with an actual counting of the number of pinning sites between the vortex and the lattice, taken to have a bcc structure. Indeed, considering an initial position (Fig. 3a) and all the possible parallel translations of the vortex line with a given orientation (θ,φ) (e.g. Fig. 3b and 3c), it is possible to obtain graphs of the number of pinning sites with respect to vortex translations (Fig. 4). Furthermore, studying those graphs for some non-aligned cases it has been possible to understand that the configurations characterized by the minimum number of pinning sites are very few, thus the vortex can migrate just avoiding them. In conclusion, we have supposed that if the vortex can reach the configurations with an average energy (and so an average number of pinning sites), it can have a sufficient number of configurations available to be considered free. On the other hand, as expected, it is bound when the number of pinning sites is maximum. Thus it is possible to rewrite the equation (4) as

$$F_L(\theta,\varphi) \propto \frac{E_p\left(N_{av}(\theta,\varphi) - N_{max}(\theta,\varphi)\right)}{\frac{1}{2}D_{max}(\theta,\varphi)} \propto E_p\left(N_{av}(\theta,\varphi) - N_{max}(\theta,\varphi)\right)\sqrt{B_{max}(\theta,\varphi)} \tag{5}$$

where we have considered that the mean distance between the bound and the free configuration can be taken as equal to the half of the average distance between the bound configurations

$$D(\theta,\varphi) \approx \frac{1}{2}D_{max}(\theta,\varphi) = \frac{1}{2}n_{max}^{-1/2}(\theta,\varphi) \propto B_{max}^{-1/2}(\theta,\varphi) \tag{6}$$

here n_{max} and B are respectively the mean density and the number of bound configurations.

Region	ρ	Element	R_{ws}	R_N	E_p		ξ	
					$\beta=1$	$\beta=3$	$\beta=1$	$\beta=3$
1	1.5×10^{12}	^{320}Zr	44.0	6.0	2.6	0.2	6.7	20.0
2	9.6×10^{12}	^{1100}Sn	35.5	6.7	1.6	0.3	4.3	13.0
3	3.4×10^{13}	^{1800}Sn	27.0	7.3	−5.2	−2.7	5.2	15.4
4	7.8×10^{13}	^{1500}Zr	19.4	6.7	−5.1	−0.7	11.2	33.5
5	1.3×10^{14}	^{982}Ge	13.8	5.2	−0.4	−0.0	38.8	116.4

Table 1. Quantities used in the calculations. The density (ρ) is given in $g\ cm^{-3}$, the pinning energy per site (E_p) is given in MeV, while the Wigner-Seitz radius (R_{ws}), the nuclear radius (R_N) and the coherence length (ξ) are given in fm.

The previous equations refer to an attractive interaction between the vortex and the nuclei of the inner crust, but depending on the matter density this interaction can be also repulsive [6, 7]. In the literature the pinning interaction is called Nuclear or Interstitial respectively, when it is attractive or repulsive.

In the interstitial pinning scenario the calculation of the pinning force is very similar to that described for the nuclear pinning regions (5). The only important difference is in the pinning energy per site: since the interaction is repulsive, the pinning energy per site is positive. Therefore, the vortex will be bound to the configurations characterized by a minimum number of pinning sites (those colored in white in figure 4). On the other hand, as in the nuclear pinning case, the vortex can be considered free if it can reach the configurations with an average number of pinning sites. In conclusion, in the interstitial pinning regions

$$F_L(\theta,\varphi) \propto E_p\left(N_{av}(\theta,\varphi) - N_{min}(\theta,\varphi)\right)\sqrt{B_{min}(\theta,\varphi)} \quad (7)$$

3. Results

The results that will be presented are obtained using different quantities from the literature (Tab. 1). Some of these concern the lattice composition (the elements that compose the cells, the Wigner-Seitz radius and the nuclear radius) and they are obtained from [15]. The others, concerning the superfluid properties, are obtained from [7]. These last values have an important dependence from a factor β that describe the intensity of the pairing gap

$$\Delta = \frac{\Delta_0}{\beta} \quad (8)$$

where Δ_0 is the bare BCS gap. This factor is related to the polarization effects of matter on the nuclear interaction. The case $\beta = 1$ describe the non-polarized situation, while the case $\beta = 3$ describe the situation in which the effect of the polarization is maximum. When $\beta = 1$ the mean pairing gap has a maximum of about 3 MeV, which corresponds to the strong pairing scenario, while when $\beta = 3$ the mean pairing gap has a maximum of about 1 MeV, as usually assumed in the weak pairing scenario.

The study presented here is done considering all the densities reported in table 1. Moreover our estimate of the maximum length for vortex rigidity is quite crude, so that it is important to consider some different vortex lengths in order to cover a greater range of possibilities ($100\ R_{ws} < L < 3000\ R_{ws}$).

The pinning forces per unit length are reported in figure 5. Here it is possible to see that there is a strong dependence on the maximum vortex length assumed in the calculation of the number of pinning sites. This behavior give a verification of Jones' hypothesis [10, 11, 12] that

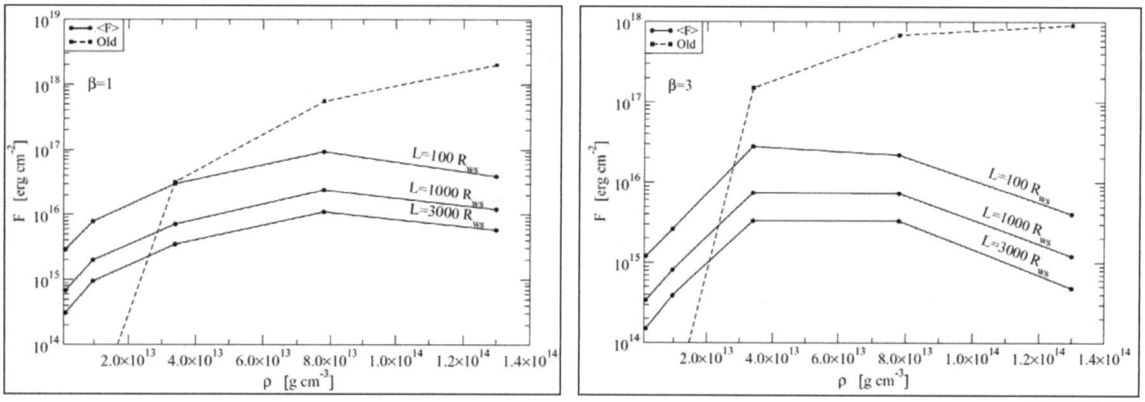

Figure 5. Pinning force per unit length as a function of the density of matter in the inner crust of a neutron star. We show both the strong pairing (left) and the weak pairing (right) scenario. The straight lines represent the pinning force per unit length calculated using three different vortex length, $L = 100\ R_{ws}$, $L = 1000\ R_{ws}$ and $L = 3000\ R_{ws}$, while the dashed lines represent the aligned case, as it is described in the literature. The quantities are given in *cgs* unit.

pinning goes to zero for infinite vortex rigidity. However, using a realistic maximum vortex length the pinning force is still not negligible.

Moreover a comparison between our results and those found in the literature (Fig. 5) shows that the maximum pinning forces per unit length obtained in this work are two or three orders of magnitude lower than those found for an aligned vortex. Thus, they are of the same order of magnitude as what has been suggested to explain the real behavior of a pulsar glitch ($\langle F \rangle \approx 10^{15} erg\ cm^{-2}$) [16]. Furthermore the pinning forces obtained with our model survive in a larger range of densities. Indeed in the presented calculations we have found that there is a considerable contribution to the pinning force also in the regions characterized by interstitial pinning, while in the same regions the existing calculations considered the vortex as free.

4. Conclusions

In this proceeding we have presented a quantitative and realistic calculation of the pinning force per unit length.

With this study it has been possible to verify the strong dependence of the pinning force on the length of vortex assumed as rigid vortex [10, 11, 12]. Indeed we find a decrease of the pinning force as the vortex length increases. On the other hand it has been possible also to disprove the conclusions given by Jones. As a matter of fact, considering a realistic portion of vortex it has been found that the pinning forces per unit length are still not negligible.

Moreover the maximum pinning forces per unit length found with this model are two or three order of magnitude smaller than those reported in literature, where the vortex has been considered as aligned with the lattice. Actually, the forces found in this work are of the same order of magnitude as those suggested to explain the behavior of a real pulsar glitch. Therefore, they can be used in realistic models of vortex dynamics in order to understand what really happens in a pulsar glitch [9].

5. Bibliography

[1] M.A. Alpar, P.W. Anderson, D. Pines and J. Shaham, Ap.J. 276, 325 (1984).
[2] P.W. Anderson and N. Itoh, Nature 256, 25 (1975).
[3] P. Avogadro, F. Barranco, R.A. Broglia and E. Vigezzi, Phys.Rev. C75, 012805 (2007).
[4] P. Avogadro, F. Barranco, R.A. Broglia and E. Vigezzi, Nucl.Phys. A811, 378 (2008).

[5] R.G. Dodson, D.R.Lewis and P.M. McCulloch, Ap&SS 308, 585 (2007).
[6] P. Donati and P.M. Pizzochero, Nucl.Phys. A742, 363 (2004).
[7] P. Donati and P.M. Pizzochero, Phys.Lett. B640, 74 (2006).
[8] R.I. Epstein and G. Baym, Ap.J. 328, 680 (1988).
[9] B. Haskell, P.M. Pizzochero and T. Sidery, arXiv:1107.5295 (2011), to appear in MNRAS.
[10] P.B. Jones, Ap.J 373, 208 (1991).
[11] P.B. Jones, Phys.Rev.Lett. 79, 792 (1997).
[12] P.B. Jones, MNRAS 296, 217 (1998).
[13] B.K. Link and R.I. Epstein, Ap.J. 373, 592 (1991).
[14] J. Middleditch, F.E. Marshall, Q.D. Wang, E.V. Gotthelf and W. Zhang, Ap.J. 625, 1531 (2006).
[15] J.N. Negele and D. Vautherin, Nucl.Phys. A207, 298 (1973).
[16] P.M. Pizzochero, arXiv:1105.0156v2 (2011), to appear in Ap.J Letters.

Journal of Physics: Conference Series **342** (2012) 012005 doi:10.1088/1742-6596/342/1/012005

Structure and Shear Modulus of the Neutron Star Crust

Joseph Hughto[1]

[1] Department of Astronomy, Indiana University, 727 E 3rd Street, Bloomington, IN 47405

E-mail: jhughto@astro.indiana.edu

Abstract. A thorough understanding of the neutron star crust is necessary to describe some observations of transient events such as quasi-periodic oscillations after magnetar giant flares and cooling following outbursts in low-mass X-ray binaries. I perform molecular dynamics simulations to model the crystalline structure of the crust and also compute shear modulus and liquid-solid phase diagrams.

1. Introduction

The structure of the neutron star(NS) crust can have important effects on observations of transient events such as quasiperiodic oscillations after magnetar giant flares[1] and cooling following an accretion-powered outburst in low-mass X-ray binaries[2]. The first question that needs to be asked when studying the crust is whether the solid crust is crystalline or amorphous. This difference can greatly affect transport properties. Ichimaru et al.[3] suggested that below the Wigner transition temperature, the OCP fluid may form an amorphous glassy state that could be long lived and thus should be taken into account when modeling NSs. Once the state of the NS crust is known, transport coefficients can be computed and these can be compared to observations.

In this paper, I will describe my molecular dynamics(MD) method in Section 2, determine whether the crust is crystalline or amorphous in Section 3, describe the shear modulus in Section 4, and conclude in Section 5.

2. Molecular Dynamics Formalism

In this section, I describe my MD formalism. For this work, I consider both a one component plasma (OCP) and mixture of two or three species. The ions interact via screened Yukawa interactions

$$v_{ij}(r) = \frac{Z_i Z_i e^2}{r} e^{-r/\lambda}, \tag{1}$$

where Z_i and Z_j are the respective charges of the two ions being considered, r is the separation between the ions, and λ is the Fermi screening length, which for cold relativistic ions is

$$\lambda^{-1} = 2k_F\sqrt{\frac{\alpha}{\pi}}. \tag{2}$$

Published under licence by IOP Publishing Ltd

Journal of Physics: Conference Series **342** (2012) 012005 doi:10.1088/1742-6596/342/1/012005

Here, the electron Fermi momentum k_F is $k_F = \left(2\pi^2 n_e\right)^{1/3}$ and α is the fine structure constant. The electron density n_e is equal to the ion charge density, $n_e = \langle Z \rangle n$, where n is the ion density and $\langle Z \rangle$ is the average charge. My simulations are classical and I have neglected the electron mass (extreme relativistic limit).

One component plasma simulations can be characterized by the Coulomb parameter $\Gamma = Z^2 e^2/aT$. I characterize my multicomponent system using an average Coulomb parameter,

$$\Gamma = \langle Z^{5/3} \rangle \Gamma_e, \tag{3}$$

where $\Gamma_e = e^2/a_e T$ with the electron sphere radius $a_e = (3/4\pi n_e)^{1/3}$.

Time can be measured in my system in units of one over the plasma frequency $\bar{\omega}_p$. For a one component plasma, the plasma frequency is

$$\omega_p = \left[\frac{4\pi Z^2 e^2 n}{M}\right]^{1/2}, \tag{4}$$

where M is the mass of the ion. For a mixture, I define an average plasma frequency and make an equation $\bar{\omega}_p = \left(4\pi \langle Z \rangle^2 e^2 n / \langle M \rangle\right)^{1/2}$. Other choices for the average plasma frequency are expected to give very similar results.

3. Crust Structure

Here I describe the theoretical evidence that supports the NS crust being a crystalline solid, rather than amorphous. In order to see if the amorphous state is stable in the NS crust, I consider amorphous initial conditions. I start with a liquid configuration of $N = 8192$ ions that is equilibrated at $\Gamma = 175$. I then quench the system instantaneously to a large Γ value by rescaling the velocities, and then evolve the system at (approximately) constant temperature until the system largely crystallizes. Table 1 lists the time needed to crystallize for different Γ values. This time increases with increasing Γ. However, this time only increases approximately linearly with Γ for $\Gamma < 1500$. This suggests that diffusion is relatively fast in the amorphous system, and that the amorphous to crystal transition does not involved a large energy barrier.. I find that the system is able to crystallize, even at $\Gamma = 1500$, where the temperature is 8.6 times lower than the melting temperature. However a final run that was quenched to $\Gamma = 1750$ was not observed to crystallize before a time $25{,}000{,}000/\omega_p$. I refer to these quenched systems as amorphous, however it may be more appropriate to call them polycrystalline because they are observed to have many small crystal domains of different orientation. These polycrystalline states are observed to undergo rapid transitions to single crystals except at the largest Γ.

To quantify the crystalline order in these simulations, I consider a metric based on bond angles[4]. Ion i is said to be bonded to ion j if it is within a distance $b = 2.44a$ that corresponds to a minimum in the radial distribution function $g(r)$. This distance is chosen to include the eight nearest neighbors in a perfect body centered cubic (bcc) lattice. Let θ_{ij} and ϕ_{ij} be the polar and azimuthal angles of radius from ion i to ion j. I calculate the spherical harmonic,

$$Q_{lm}\left(\hat{r}_{ij}\right) = Y_{lm}\left(\theta_{ij}, \phi_{ij}\right), \tag{5}$$

and average over all $\approx 14N$ bonds for a given configuration,

$$\bar{Q}_{lm} = \left\langle Q_{lm}\left(\hat{r}_{ij}\right)\right\rangle. \tag{6}$$

This quantity depends on the orientation of a crystal lattice with respect to the simulation volume. Therefore, I calculate the rotationally invariant quantity Q_l[4],

$$Q_l = \left[\frac{4\pi}{2l+1} \sum_{m=-l}^{l} \left|\bar{Q}_{lm}\right|^2\right]^{1/2}. \tag{7}$$

Table 1. Approximate time t_0 for amorphous systems to crystallize, after the systems have been instantaneously quenched from $\Gamma = 175$ to different Γ values. N is the number of ions in the system.

N	Γ	$t_0\omega_p$
8192	500	24,000
8192	600	47,000
8192	700	142,000
8192	1000	240,000
8192	1500	390,000
8192	1750	>25,000,000
27648	500	400,000
27648	1000	>6,000,000

This provides a measure of the crystalline order of a configuration. In general, Q_l is small for a liquid or amorphous configuration and Q_l is large for a perfect crystal. My calculations of Q_l, for a range of even l, show that Q_6 is most sensitive to crystalline order. I find that $Q_6 = 0.51069$ for a perfect, i.e. $T = 0$, bcc lattice, and $Q_6 = 0.57452$ for a perfect face centered cubic lattic. Note that Q_l is small for odd l. In Figure 1 I show Q_6 versus simulation time t_0. However, the amount of time necessary for Q_6 to grow can increase strongly with increasing system size N or Γ. A plateau near $Q_6 \approx 0.17$ is see for all four systems in Figure 1, which suggests a possible metastable intermediate state. The simulations with $N = 27648$ at $\Gamma = 500$ and $N = 8192$ at $\Gamma = 1500$ show a rapid rise in Q_6, near $t_0 = 4 \times 10^5/\omega_p$, during transitions to single crystals. For $N = 27648$ at $\Gamma = 1000$ and $N = 8192$ at $\Gamma = 1750$, Q_6 is increasing with time. These systems have not yet evolved into single crystals, however the continued rise of Q_6 with time strongly suggests that these sytems will evolve into single crystals at later times and are unlikely to remain amorphous.

I conclude that an amorphous solid will not form even with large amounts of supercooling, where the temperature is rapidly quenched by up to a factor of 10 below the melting temperature. Instead, diffusion is fast enough so that the system will form a regular crystal. My results stronly suggest that Coulomb solids in the interior of cold white dwarf stars and the crust of NSs will be crystalline and not amorphous. This is consistent with observations of rapid crust cooling of neutron stars following extended periods of accretion[2]. This rapid cooling implies a high crust thermal conductivity, that agrees with a the conductivity of a regular crystal, and is larger than that expected for an amorphous solid.

4. Shear Modulus

I now describe my methodology and results for the shear modulus in the NS crust. To calculate shear moduli, I follow the procedure of Ogata et al.[5]. The change in free energy with deformation δF can be expressed in terms of elastic constants c_{11}, c_{12}, and c_{44}.

$$\delta F = \frac{1}{2}\left(c_{11} - c_{12}\right) u_{ii}^2 + c_{44} u_{ik} u_{ki} \quad (i \neq k), \tag{8}$$

and u_{ik} describes the strain.

Under a deformation, the coordinates r_k of an ion get mapped to r'_i,

$$r'_i = \sum_{k=1}^{3} \left(\delta_{ik} + u_{ik}\right) r_k. \tag{9}$$

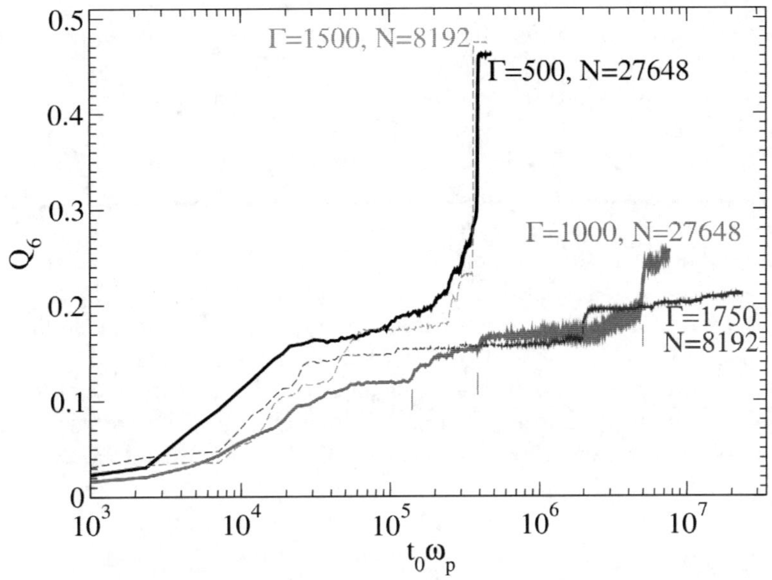

Figure 1. Bond angle metric Q_6, see Equations 5,6,7, versus simulation time t_0 for amorphous systems that were instantaneously quenched from a $\Gamma = 175$ liquid at $t_0 = 0$. The number of ions in the simulation N and Coulomb parameter Γ are indicated.

I consider six deformations D_i $(i = 1 \dots 6)$ that conserve the volume to order ϵ^2.

$$D_1: \quad u_{xx} = \epsilon + \frac{3}{4}\epsilon^2, \quad u_{yy} = u_{zz} = -\frac{\epsilon}{2} \tag{10}$$

$$D_2: \quad u_{yy} = \epsilon + \frac{3}{4}\epsilon^2, \quad u_{xx} = u_{zz} = -\frac{\epsilon}{2} \tag{11}$$

$$D_3: \quad u_{zz} = \epsilon + \frac{3}{4}\epsilon^2, \quad u_{xx} = u_{yy} = -\frac{\epsilon}{2} \tag{12}$$

$$D_4: \quad u_{xy} = u_{yx} = \frac{\epsilon}{2}, \quad u_{zz} = \frac{\epsilon^2}{4} \tag{13}$$

$$D_5: \quad u_{yz} = u_{zy} = \frac{\epsilon}{2}, \quad u_{xx} = \frac{\epsilon^2}{4} \tag{14}$$

$$D_6: \quad u_{zx} = u_{xz} = \frac{\epsilon}{2}, \quad u_{yy} = \frac{\epsilon^2}{4} \tag{15}$$

For each defomation D_m I calculate a corresponding expectation value f_m $(m = 1 \dots 6)$,

$$f_m = \frac{1}{V}\left\{\left\langle \frac{d^2V_{tot}}{d\epsilon^2} \right\rangle - \frac{1}{T}\left[\left\langle \left(\frac{dV_{tot}}{d\epsilon}\right)^2 \right\rangle - \left\langle \frac{dV_{tot}}{d\epsilon} \right\rangle^2 \right]\right\}, \tag{16}$$

where V is the system volume and V_{tot} is the total potential under deformation D_m. At zero temperature, this reduces to $f_m = (d^2V_{tot}/d\epsilon^2)/V$.

For a bcc crystal, one has[5] $f_1 = f_2 = f_3 = 3b_{11} = 3\,(c_{11} - c_{12})$ and $f_4 = f_5 = f_6 = c_{44}$. Here, c_{11}, c_{12}, and c_{44} are elastic constants. In practice I calculate all six f_m independently and average to determine b_{11} and c_{44}. The angle averaged shear modulus is[5],

$$\mu_{eff} = (2b_{11} + 3c_{44})/5. \tag{17}$$

If the NS crust involves many crystal domains of random orientation, then μ_{eff} is the appropriate elastic constant to determine the speed of shear waves.

For simplicity, I work at a density $n = 7.18 \times 10^{-5}$ fm^{-3} and $Z = 29.4$. My results can be scaled to other densities at a given value of Γ. My results can also be approximately scaled to other values of Z, at fixed Γ. This is because, although the ratio of λ/a changes with Z, this change in screening has only a small effect on the shear modulus. I evolve the system with the velocity Verlet algorithm[6] using a time step $\delta t = 25$ fm/c. Starting from $T = 0$ and a perfect bcc lattice, I increase the temperature to $T = 0.1$ MeV and evolve the system for typically 100000 MD steps (2.5×10^6 fm/c) to reach thermal equilibrium. Next, I evolve for a further 250000 MD steps (6.25×10^6 fm/c) storing configurations for later calculations of elastic constants. The temperature is then raised by 0.1 MeV and the process is repeated. I keep the system at a fixed temperature (approximately) by periodically rescaling the velocities. These MD simulations are done in an undistorted cubic box using periodic boundary conditions.

I calculate f_m by averaging over 1000 configurations, each separated by 250 MD steps. To minimize finite size effects, I calculate V_{tot} by summing over all 27 nearest periodic images. Thus ion i is assumed to interact not only with ion j at its original position, but also with 26 more images of j where the x, y, and z coordinates are independently shifted by 0, $+l$, or $-l$, with l being the box size. The derivatives in Equation 16 are approximated using a five point numerical formula. I note that the MD trajectories have been calculated using periodic distances (involving only the single nearest periodic image of a given ion) to save time, while the derivatives have been calculated by summing over 27 images to minimize finite size effects.

Table 2 presents results for simulations using $N = 3456$ ions. Statistical errors only are indicated in partheses. I caution that b_{11} may have significant errors from finite size and other systematic effects. I fit the values of μ_{eff} in Table 2 with a simple analytic formula that is valid for all $\Gamma \geq 175$,

$$\mu_{eff} \approx \left(0.1106 - \frac{28.7}{\Gamma^{1.3}}\right)\left(n\frac{Z^2e^2}{a}\right) \tag{18}$$

This fit has an error $\leq 2\%$.

Table 2. Shear moduli for MD simulations with $N = 3456$ ions in units of nZ^2e^2/a.

Γ	b_{11}	c_{44}	μ_{eff}
∞	0.0220	0.1699	0.1107
834	0.0209(2)	0.1617(3)	0.1054(2)
417	0.0194(2)	0.1517(3)	0.0988(2)
278	0.0202(4)	0.1410(5)	0.0927(3)
200	0.0154(5)	0.1253(10)	0.0813(6)
175	0.0158(8)	0.1152(10)	0.0755(6)

To study finite size effects, I have performed additional simulations with larger and smallers systems of sizes $N = 9826$ and $N = 1024$, respectively. Since the μ_{eff} is about 1% smaller for $N = 9826$ compared to $N = 3456$, I estimate finite size effects to be of order 1%. Figure 2 plots these results for μ_{eff} and also shows the Monte Carlo results of Ogata et al.[5]. These results are about 10% larger than my results at large Γ and have much larger statistical errors.

Ogata et al. neglect electron screening $\lambda \to \infty$. At zero temperature, I have performed calculations for larger values of λ and extrapolated to $\lambda \to \infty$. My extrapolated results are

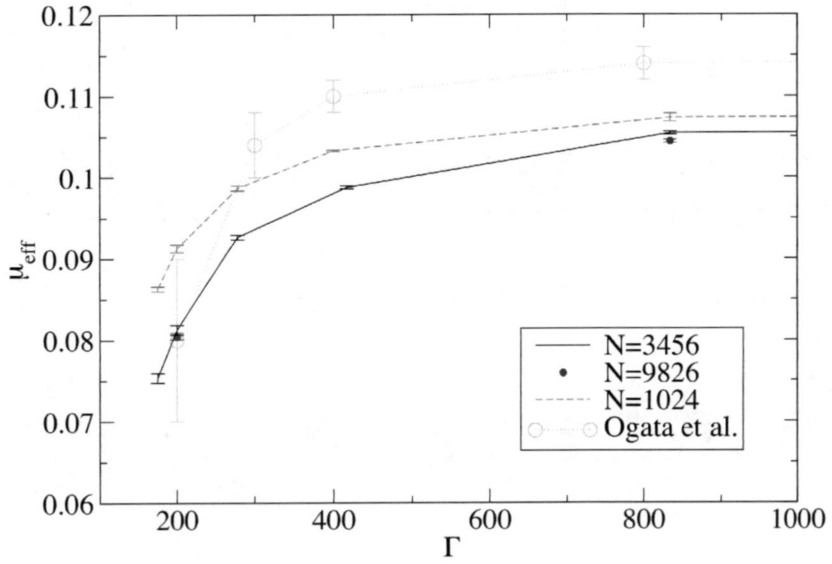

Figure 2. Angle averaged shear modulus μ_{eff} versus Coulomb parameter Γ for MD simulations involving $N = 1024$, 3456, and 9826 ions. Also shown are Monte Carlo results from Ogata et al.[5] that omit electron screening.

consistant with Ogata et al. Therefore, I conclude that electron screening reduces μ_{eff} be about 10%. The speed of shear waves is proportional to the square root of μ_{eff}, so electron screening reduces shear speed by about 5%. This will slightly lower the frequency of torsional oscillations of neutron star crusts.

5. Conclusion

I have shown that the NS crust is most likely crystalline. The fact that amorphous systems are unstable to spontaneous crystallization suggests that macroscopic sections of the NS crust should be a regular crystal. Observational results also suggest that the crust is crystalline since the low thermal conductivity of an amorphous solid cannot explain observations.

Since the crust is most likely crystalline, the shear modulus can be computed including the effect of electron screening that was neglected in previous calculations. Including electron screening reduces the shear modulus by about 10% which implies a corresponding reduction in shear speed of 5%. This difference should be observable with high quality data of quasi-periodic oscillations after magnetar giant flares.

References

[1] Steiner A W and Watts A L 2009 *Physical Review Letters* **103** (*Preprint* 0902.1683)
[2] Shternin P S, Yakovlev D G, Haensel P and Potekhin A Y 2007 *MNRAS* **382** (*Preprint* 0708.0086)
[3] Ichimaru S, Iyetomi H, Mitake S and Itoh N 1983 *Ap. J. Letters* **265** L83–L86
[4] Steinhardt P J, Nelson D R and Ronchetti M 1983 *Phys. Rev. B* **28** 784–805
[5] Ogata S and Ichimaru S 1990 *Phys. Rev. A* **42** 4867–4870
[6] Verlet L 1967 *Physical Review* **159** 98–103

Journal of Physics: Conference Series **342** (2012) 012006 doi:10.1088/1742-6596/342/1/012006

Effect of hyperonic three-body forces on the maximum mass of neutron stars

D. Logoteta [1], I. Vidaña[1], C. Providência [1], A. Polls [2] and I. Bombaci [3]

[1]Centro de Física Computacional, Department of Physics, University of Coimbra, PT-3004-516, Coimbra, Portugal

[2] Departament d'Estructura i Constituents de la Matèria and Institut de Ciències del Cosmos, Universitat de Barcelona, Avda. Diagonal 647, E-08028 Barcelona, Spain

[3] Dipartimento di Fisica "E. Fermi", Università di Pisa, and INFN, Sezione di Pisa, Largo B. Pontecorvo 3, I-56127 Pisa, Italy

Abstract.

A model based on a microscopic Brueckner–Hartree–Fock approach of hyperonic matter supplemented with additional simple phenomenological density-dependent contact terms is employed to estimate the effect of hyperonic three-body forces on the maximum mass of neutron stars. Our results show that although hyperonic three-body forces can reconcile the maximum mass of hyperonic stars with the current limit of $1.4 - 1.5M_{\odot}$, they are unable to provide the repulsion needed to make the maximum mass compatible with the observation of massive neutron stars, such as the recent measurements of the unusually high masses of the millisecond pulsars PSR J1614-2230 ($1.97 \pm 0.04M_{\odot}$) and PSR J1903+0327 ($1.667 \pm 0.021M_{\odot}$).

1. Introduction

Neutron stars are the remnants of the gravitational collapse of massive stars during a Type II, Ib or Ic supernova explosion. Their masses and radii are typically of the order of $1 - 2M_{\odot}$ ($M_{\odot} \simeq 2 \times 10^{33}$ g is the mass of the Sun) and $10 - 12$ km, respectively. With central densities in the range of $4 - 8$ times the normal nuclear matter saturation density, $\epsilon_0 \sim 2.7 \times 10^{14}$ g/cm^3 ($\rho_0 \sim 0.16$ fm^{-3}), neutron stars are most likely among the densest objects in the universe [1]. Nowadays, it is still an open question which is the true nature of neutron stars. Traditionally the core of neutron stars has been modeled as an uniform fluid of neutron rich nuclear matter in equilibrium with respect to the weak interaction (β-stable nuclear matter). Nevertheless, due to the large value of the density, new hadronic degrees of freedom are expected to appear in addition to nucleons. Hyperons, baryons with a strangeness content, are an example of these degrees of freedom. Since the pioneer work of Ambartsumyan and Saakyan [2] the presence of hyperons in neutron stars has been studied by many authors using either phenomenological [3, 4, 5, 6] or microscopic [7, 8, 9, 10] approaches. Hyperons may appear in the inner core of neutron stars at densities of about $2 - 3\rho_0$. At such densities, the nucleon chemical potential is large enough to make the conversion of nucleons into hyperons energetically favorable. This conversion relieves the Fermi pressure exerted by the baryons, and makes the equation of state (EoS) softer. As a consequence, the maximum mass of the star is substantially reduced to values that, in microscopic calculations [7, 8, 9, 10], can be even below the "canonical" one of

$1.4-1.5M_\odot$ inferred from precise neutron star mass determinations [11]. It has been suggested, see *e.g.* Ref. [12], that a possible solution to this problem is the inclusion of three-body forces (TBF) involving one or more hyperons (*i.e.*, nucleon-nucleon-hyperon (NNY), nucleon-hyperon-hyperon (NYY), and hyperon-hyperon-hyperon (YYY)). These forces could eventually provide the additional repulsion needed to make the EoS stiffer and, therefore, the maximum mass compatible with the current observational limits. In this letter, we want to establish numerical lower and upper limits to the effect of these forces on the maximum mass of neutron stars that can guide more sophisticated calculations in the future. To such end, we use a model based on a microscopic Brueckner–Hartree–Fock (BHF) approach of hyperonic matter [7, 8, 9, 13] supplemented with additional simple phenomenological density-dependent contact terms that account for the effect of nucleonic and hyperonic TBF.

2. Formalism

Our calculation starts by constructing all the baryon-baryon G matrices, which describe the interaction between two baryons in the presence of a surrounding medium. The G matrices are obtained by solving the coupled-channel Bethe–Goldstone equation, written schematically as

$$G(\omega)_{B_1B_2,B_3B_4} = V_{B_1B_2,B_3B_4} + \sum_{B_iB_j} V_{B_1B_2,B_iB_j} \times \frac{Q_{B_iB_j}}{\omega - E_{B_i} - E_{B_j} + i\eta} G(\omega)_{B_iB_j,B_3B_4} \,, \quad (1)$$

where the first (last) two subindices indicate the initial (final) two-baryon states compatible with a given value S of the strangeness, namely nucleon-nucleon (NN) for $S=0$ and hyperon-nucleon (YN) for $S=-1$; $V_{B_1B_2,B_3B_4}$ is the bare baryon-baryon interaction (NN→NN, ΛN→ΛN, ΛN→ΣN or ΣN→ΣN); $Q_{B_iB_j}$ is the Pauli operator, that prevents the intermediate baryons B_i and B_j from being scattered to states below their respective Fermi momenta; and the starting energy ω corresponds to the sum of the nonrelativistic single-particle energies of the interacting baryons.

The single-particle energy of a baryon B_i is given by

$$E_{B_i}(\vec{k}) = M_{B_i} + \frac{\hbar^2 k^2}{2M_{B_i}} + \mathrm{Re}[U_{B_i}(\vec{k})] \,. \quad (2)$$

Here M_{B_i} denotes the rest mass of the baryon, and the single-particle potential U_{B_i} represents the average field "felt" by the baryon owing to its interaction with the other baryons of the medium. In the BHF approximation, U_{B_i} is calculated through the "on-shell energy" G-matrix, and is given by

$$U_{B_i}(\vec{k}) = \sum_{B_j}\sum_{\vec{k'}} n_{B_j}(|\vec{k'}|) \times \langle \vec{k}\vec{k'}|G(E_{B_i}(\vec{k}) + E_{B_j}(\vec{k'}))_{B_iB_j,B_iB_j}|\vec{k}\vec{k'}\rangle_{\mathcal{A}} \,, \quad (3)$$

where $n_{B_j}(|\vec{k}|)$ is the occupation number of the species B_j, and the index $\mathcal{A}$ indicates that the matrix elements are properly antisymmetrized when baryons B_i and B_j belong to the same isomultiplet. We note here that the so-called continuous prescription has been adopted for the single-particle potentials when solving the Bethe–Goldstone equation, since, as shown by the authors of Ref. [14], the contribution to the energy per particle from three-hole line diagrams is minimized in this prescription. All the calculations carried out in this letter have been performed with the realistic Argonne V18 [15] NN force and the Nijmegen soft-core NSC89

Journal of Physics: Conference Series **342** (2012) 012006 doi:10.1088/1742-6596/342/1/012006

[16] YN interaction. We recall that the NSC89 potential does not contain hyperon-hyperon (YY) components.

Once a self-consistent solution of Eqs. (1)–(3) is obtained, the energy density can be calculated in the BHF approximation as

$$\epsilon_{BHF} = \frac{1}{\mathcal{V}} \sum_{B_i} \sum_{\vec{k}} n_{B_i}(|\vec{k}|) \left[M_{B_i} + \frac{\hbar^2 k^2}{2M_{B_i}} + \frac{1}{2} U_{B_i}(\vec{k}) \right] . \quad (4)$$

TBF can be introduced in the BHF approach by adding effective density-dependent two-body forces to the baryon-baryon interactions V when solving the Bethe–Goldstone equation for the G-matrices. These effective forces are obtained by averaging genuine baryon-baryon-baryon TBF, $W_3(\vec{r}_i, \vec{r}_j, \vec{r}_k)$, over the third baryon coordinates [17],

$$V^{eff}_{B_iB_j}(\vec{r}_{ij}) = \int W_3(\vec{r}_i, \vec{r}_j, \vec{r}_k) n(\vec{r}_i, \vec{r}_j, \vec{r}_k) d\vec{r}_k , \quad (5)$$

where $\vec{r}_{ij}$ is the relative coordinate of baryons B_i and B_j, and $n(\vec{r}_i, \vec{r}_j, \vec{r}_k)$ is an appropriate three-body correlation function. This procedure has been the usual one to introduce nucleonic TBF in BHF and other non-relativistic many-body approaches of nuclear matter (see *e.g.*, Ref. [18]). Nevertheless, since there is an almost complete lack of experimental data, and few theoretical studies [19, 20, 21, 22, 23, 24, 25, 26, 27], on the genuine hyperonic TBF, instead of using the average procedure described, in this letter we adopt an alternative strategy. First, we construct, as described above, the hyperonic matter EoS within the BHF approach using only two-body NN and YN forces, and then, we add to it simple phenomenological density-dependent contact terms that account for the effect of both nucleonic and hyperonic TBF. The reader could think that in doing that the predictive power of our model is partially lost. However, we would like to stress once more that the aim of this work is to perform a simple estimation of the effect of hyperonic TBF on the maximum mass of neutron stars that can guide and motivate more realistic and sophisticated studies of such forces and their effects. We want also to point out that by using a simple parametrization of the hyperonic TBF we can easily explore whether or not such forces can make the EoS stiff enough so that the maximum mass of hyperonic stars can be reconciled with the observational limits. Following Balberg and Gal (see Eqs. (2) and (3) of Ref. [6]), we assume these density-dependent contact terms to have a Skyrme-like form

$$\epsilon_{xy}(\rho_x, \rho_y) = a_{xy}\rho_x\rho_y + b_{xy}\rho_x\rho_y \left(\frac{\rho_x^{\gamma_{xy}} + \rho_y^{\gamma_{xy}}}{\rho_x + \rho_y} \right) , \quad (6)$$

where x and y denote any baryon species. Note that, to be precise only the second term actually accounts for the effect of three- and multi-body forces which are responsible for the repulsion that dominates at high densities. The first one is in fact a two-body term that modifies slightly the BHF part of the energy density by yielding additional attraction at low densities. It can be considered as a rearrangement of the two-body contribution due to the inclusion of TBF. It mimics the modification that the effective density-dependent two-body force, $V^{eff}_{B_iB_j}$, would introduce on the G-matrices. Then, we have the following contact terms contributing to the energy density

$$\begin{aligned} \epsilon_{CT} &= a_{NN}\rho_N^2 + b_{NN}\rho_N^{\gamma_{NN}+1} \\ &+ a_{\Lambda N}\rho_\Lambda\rho_N + b_{\Lambda N}\rho_\Lambda\rho_N \left(\frac{\rho_\Lambda^{\gamma_{\Lambda N}} + \rho_N^{\gamma_{\Lambda N}}}{\rho_\Lambda + \rho_N} \right) \\ &+ a_{\Sigma N}\rho_\Sigma\rho_N + b_{\Sigma N}\rho_\Sigma\rho_N \left(\frac{\rho_\Sigma^{\gamma_{\Sigma N}} + \rho_N^{\gamma_{\Sigma N}}}{\rho_\Sigma + \rho_N} \right) , \end{aligned} \quad (7)$$

Journal of Physics: Conference Series **342** (2012) 012006 doi:10.1088/1742-6596/342/1/012006

γ_{NN}	a_{NN} [MeV fm^3]	b_{NN} [MeV fm$^{3\gamma_{NN}}$]	K_∞ [MeV]
2	−33.44	213.02	211
2.5	−22.08	355.03	236
3	−16.40	665.68	260
3.5	−12.99	1331.36	285

Table 1. Values of the parameters a_{NN}, b_{NN}, γ_{NN}, and the incompressibility coefficient K_∞.

where $\rho_N = \rho_n + \rho_p$, $\rho_\Sigma = \rho_{\Sigma^-} + \rho_{\Sigma^0} + \rho_{\Sigma^+}$, and we assume charge independence. We note that the first two terms mime the contribution of the effective two-body interation, V_{NN}^{eff}, obtained when the forces NNN and NNY are averaged over the coordinates of one of the nucleons and the hyperon, respectively (see Eq. (5)). Similarly, the remaining terms mimic the contributions of the effective V_{YN}^{eff} forces built from the average of the NNY and NYY ones over the coordinates of the appropriate nucleons and hyperons. Note that the NYY forces can also contribute, in addition to the YYY ones, to the effective density-dependent YY forces if the average is taken over the nucleon coordinates. For consistency with the use of the NSC89 YN potential, which as it is said, does not describe any YY interaction, none of the terms of Eq. (7) mimic neither this other contribution of NYY, nor the ones of YYY. The complete baryonic energy density is then simply obtained by adding ϵ_{CT} to the BHF energy density of Eq. (4).

The parameters a_{NN}, b_{NN} and γ_{NN} are fitted to reproduce the density, ρ_0, energy per particle, E/A, and incompressibility coefficient, K_∞, of symmetric nuclear matter at saturation. We choose the saturation point to be at $\rho_0 = 0.16$ fm^{-3} and $E/A = -16$ MeV. Due to the uncertainty still existing on the precise value of K_∞, we consider four values of γ_{NN} between 2 and 3.5, giving K_∞ in the range 211 − 285 MeV, compatible with the value of $K_\infty = 240 \pm 40$ MeV supported nowadays by experimental data [28]. The values of a_{NN}, b_{NN}, γ_{NN} and K_∞ are given in Table 1. We note that due to the charge independence assumed in Eq. (7) neither the symmetry energy, $E_{sym}(\rho)$, nor its derivative parameter $L = 3\rho_0 \partial E_{sym}(\rho)/\partial\rho|_{\rho_0}$ will be affected by the inclusion of the contact terms. Their values at $\rho_0 = 0.16$ fm^{-3} are 28.3 and 46.6 MeV, respectively.

In order to reduce the number of parameters, in this exploratory work, we assume for simplicity that TBF involving Λ and Σ hyperons are the same. Therefore, we take $a_{\Lambda N} = a_{\Sigma N}, b_{\Lambda N} = b_{\Sigma N}$ and $\gamma_{\Lambda N} = \gamma_{\Sigma N}$. The parameters a_{YN} and b_{YN} ($Y = \Lambda, \Sigma$) are considered free. We take them to be a fraction x of the parameters a_{NN} and b_{NN} ($a_{YN} = x\,a_{NN}$ and $b_{YN} = x\,b_{NN}$ with $x = 0, 1/3, 2/3$ and 1) in order to explore different strengths of the hyperonic TBF. There is no physical reason for which these parameters should scale with the same factor x with respect to their nucleonic counterparts. However, we have checked that allowing different scaling factors for a_{YN} and b_{YN}, we always obtain maximum masses within the range of values shown in Table 2. As an example, we note that for γ_{NN}=3.5 we obtain $M_{max} = 1.61 M_\odot$ for $a_{YN} = a_{NN}/3$ and $b_{YN} = b_{NN}$, which is very similar to the value of $1.60 M_\odot$ shown in Table 2 for $x = 1$. In view of this, and for simplicity, we keep the same scaling factor x for both a_{YN} and b_{YN}. Finally, the value of −28 MeV for the binding energy of a single Λ in nuclear matter at saturation, extracted from the extrapolation of hypernuclear data [29], is then used to determine the parameter γ_{YN} through the relation

$$\left(\frac{B}{A}\right)_\Lambda = -28 \text{ MeV} = U_\Lambda(k=0) + a_{YN}\rho_0 + b_{YN}\rho_0^{\gamma_{YN}} , \tag{8}$$

γ_{NN}	x	γ_{YN}	M_{max}	ρ_c	v_s
	0	-	1.27 (2.22)	1.35 (1.07)	0.46 (1.03)
	1/3	1.49	1.33	1.33	0.48
2	2/3	1.69	1.38	1.29	0.52
	1	1.77	1.41	1.24	0.54
	0	-	1.29 (2.46)	1.19 (0.92)	0.43 (1.17)
	1/3	1.84	1.38	1.16	0.49
2.5	2/3	2.08	1.44	1.12	0.54
	1	2.19	1.48	1.09	0.56
	0	-	1.34 (2.72)	0.98 (0.79)	0.40 (1.34)
	1/3	2.23	1.45	0.97	0.50
3	2/3	2.49	1.50	0.94	0.55
	1	2.62	1.54	0.90	0.58
	0	-	1.38 (2.97)	0.87 (0.69)	0.38 (1.47)
	1/3	2.63	1.51	0.86	0.51
3.5	2/3	2.91	1.56	0.83	0.56
	1	3.05	1.60	0.80	0.59

Table 2. Maximum neutron star mass, central baryon number density and central speed of sound for different values of the contact term parameters. The results for $x = 0$ correspond to the case when only nucleonic TBF are considered. In brakets are given the corresponding values when the presence of hyperons is neglected. Masses are given in $M_\odot$ whereas the central baryon density, ρ_c, is given in fm^{-3} and the central speed of sound is given in units of c.

where $U_\Lambda(k = 0)$ (-30.8 MeV in our calculation) is the BHF single-particle potential of the Λ (see Eq. (3)) at zero momentum, and the last two terms account for the contact term contribution.

Once the total energy density $\epsilon = \epsilon_{BHF}+\epsilon_{CT}+\epsilon_L$ (ϵ_L being the contribution of noninteracting leptons), is known, the composition and the EoS of neutron star matter can be obtained from the requirement of equilibrium under weak interaction processes, $\mu_i = b_i\mu_n - q_i\mu_e$ (b_i and q_i denoting the baryon number and charge of species i) and electric charge neutrality, $\sum_i q_i\rho_i = 0$. The chemical potentials of the various species and the pressure are computed from the usual thermodynamical relations, $\mu_i = \partial\epsilon/\partial\rho_i$ and $P = \rho^2\partial(\epsilon/\rho)/\partial\rho$. Finally, knowing the EoS, the equilibrium configurations of static neutron stars are obtained by solving the well-known Tolman–Oppenheimer–Volkoff (TOV) equations [1].

3. Results and conclusions

Table 2 shows the maximum neutron star mass, central baryon number density and central speed of sound for different values of the contact term parameters. The range of variation of these parameters allows us to explore different EoS and establish a lower and an upper limit to the effect of hyperonic TBF on the maximum mass. One can get an idea of the relative importance of these forces on the EoS (and, therefore, in some way also of the relative weight of the phenomenological part of our calculation with respect to the microscopic one), by evaluating the ratio between the last four terms of Eq. (7), and the sum $\sum_{B_i}\sum_{\vec{k}} n_{B_i}(|\vec{k}|)U_{B_i}(\vec{k})/(2\mathcal{V})$ (see Eq. (4)). We find that, in average, this ratio is smaller than 0.2 for $\rho < 4\rho_0$, it ranges between 0.2 and 0.5 for densities up to $\sim 5\rho_0$, and it is larger than 1 for $\rho > 6\rho_0$, clearly showing that the relative importance of hyperonic TBF increases for larger densities. The results for $x = 0$ correspond to the case when only nucleonic TBF are considered (*i.e.*, $a_{YN} = b_{YN} = 0$). In brakets are given the corresponding values when the presence of hyperons is neglected in the

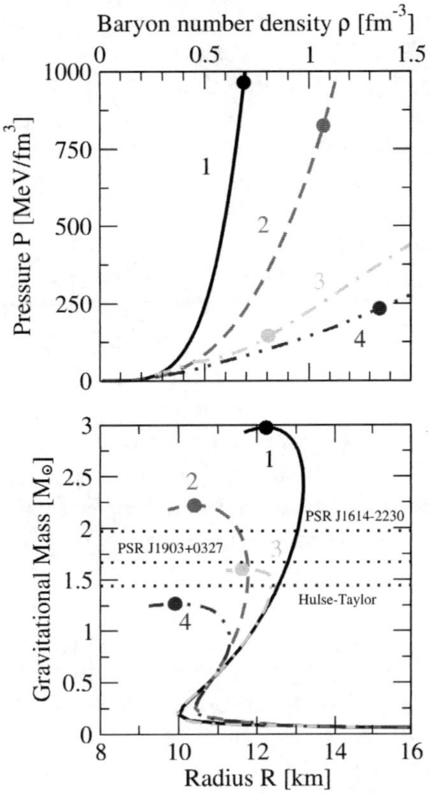

Figure 1. (Color online) Upper panel: β-stable matter EoS. Lower panel: Mass-radius relation for different EoS. Circles indicate the central baryon number density, central pressure, mass and radius of the maximum mass stellar configuration. Horizontal lines show the masses of the pulsars PSR J1614-2230 [30], PSR J1903+0327 [31] and the Hulse–Taylor one [32]. See the text for details.

EoS. Note that in this case the resulting maximum mass is relatively large, ranging from $2.22M_\odot$ for $\gamma_{NN} = 2$ to $2.97M_\odot$ for $\gamma_{NN} = 3.5$. The presence of hyperons induces here a reduction of the mass to values in the interval $1.27 - 1.38M_\odot$, well below the limit of $1.4 - 1.5M_\odot$. Note that the range of variation of the maximum mass is about $0.11M_\odot$ in this case, compared to a range of $\sim 0.75M_\odot$ when hyperons are absent. This is a consequence, as already pointed out in Ref. [9], of a strong compensation mechanism caused by the appearance of hyperons which makes the maximum mass quite insensitive to the pure nucleonic part of the EoS. As expected, the central density decreases when increasing the effect of three-body forces (the pressure is larger and consequently the object is less compact), but at the same time the speed of sound increases, because the EoS is stiffer. We note that when the presence of hyperons is neglected the EoS is always supraluminical. This is not surprising, since our approach is a non-relativistic one, and causality is, therefore, not guaranteed. However, note that as soon as hyperons are present in matter, the softening of the EoS induced by their presence is such that in these cases the EoS remains always causal. It is clear from Table 2 that hyperonic TBF provide additional repulsion making the EoS stiffer, and the maximum mass larger. For a fixed value of the exponent γ_{NN} the maximum mass increases when increasing x (*i.e.*, γ_{YN}). This is an expected result, since by increasing x we are increasing the strength of the hyperonic TBF and, as a consequence, the

EoS becomes stiffer, and the maximum mass larger. We have checked that the rate of increase of M_{max} with x is slightly quadratic. The stiffer EoS including hyperonic TBF is obtained for $\gamma_{NN} = 3.5$ and $x = 1$ ($\gamma_{YN} = 3.05$), and allows for a maximum mass of about $1.60M_\odot$. We note that although the inclusion of hyperonic TBF can reconcile the maximum mass of hyperonic stars with the "canonical" value, they are, however, unable to make the maximum mass compatible with the observation of massive neutron stars, such as the recent measurements of the unusually high masses of the millisecond pulsars PSR J1614-2230 ($1.97 \pm 0.04M_\odot$) [30] and PSR J1903+0327 ($1.667 \pm 0.021M_\odot$) [31]. One could in principle increase arbitrarily the strength of the hyperonic TBF by increasing the value of the fraction x in order to get a very stiff EoS which supports such large masses. However, in our opinion values of $x > 1$ give rise to EoS unrealistically stiff, the reason being the following: we know that the strength of the two-body YN interaction is smaller than that of the NN one (*e.g.*, the single-particle potential of a Λ in symmetric nuclear matter at saturation for zero momentum is about 1/3 that of the nucleons). Therefore, it is quite natural to think that probably the strength of hyperonic TBF is either smaller or as large as the pure nucleonic ones, but not larger. Although this statement seems a bit speculative, and only a more realistic and sophisticated determination of hyperonic TBF can give a definite answer, we consider the value of $1.60M_\odot$ a reasonable upper limit for the maximum mass of neutron stars with hyperonic TBF. This upper limit could be slightly larger, not more than 5%, if we would have taken $a_{\Sigma N} \neq a_{\Lambda N}, b_{\Sigma N} \neq b_{\Lambda N}$, and $\gamma_{\Sigma N} \neq \gamma_{\Lambda N}$, and assumed a slightly repulsive single-particle potential for the Σ^- in nuclear matter at saturation density. We point out, however, that the uncertainties related with the binding-energy of the Σ^- in nuclear matter at saturation density are still very large (see *e.g.*, Refs. [33, 34, 35, 36]). Clearly, $1.27M_\odot$ is the lower limit corresponding to the softer EoS.

Fig. 1 shows as a summary the EoS (upper panel) for the stiffer and the softer pure nucleonic (curves 1 and 2) and hyperonic (curves 3 and 4) stars, and their corresponding mass-radius relation (lower panel). Curves 1 and 2 show respectively the results for $\gamma_{NN} = 3.5$ and $\gamma_{NN} = 2$ without hyperons, while curves 3 and 4 correspond to the cases $\gamma_{NN} = 3.5, x = 1$, and $\gamma_{NN} = 2, x = 0$. Circles indicate the value of the central baryon number density, central pressure, mass and radius of the maximum mass stellar configuration. The horizontal lines show the masses of the pulsars PSR J1614-2230, PSR J1903+0327 and the Hulse–Taylor one (1.4414 ± 0.0002) [32]. The strong softening of the EoS due to the presence of hyperons and the consequent reduction of the maximum mass is clearly seen. The maximum masses of hyperonic stars lay in a narrow range from 1.27 to $1.60M_\odot$, that is still compatible with the mass of Hulse-Taylor pulsar, but is well below the masses of PSR J1903+0327 and PSR J1614-2230.

Summarizing, we use a model based on a microscopic BHF approach of hyperonic matter supplemented with additional simple phenomenological density-dependent contact terms to establish numerical lower and upper limits to the effect of hyperonic TBF on the maximum mass of neutron stars. Assuming that the strength of these forces is either smaller or as large as the pure nucleonic ones, our results show that maximum masses of hyperonic stars lay in a narrow range from 1.27 to $1.60M_\odot$ which is still compatible with the "canonical" value of $1.4 - 1.5M_\odot$, but it is incompatible with the observation of massive neutron stars, such as the recent measurements of a mass of $1.97 \pm 0.04M_\odot$ for the millisecond pulsar PSR J1614-2230, and a mass of $1.667 \pm 0.021M_\odot$ for the PSR J1903+0327 one. We hope that this work may stimulate the calculation of microscopic hyperonic three body forces that can be used to test the conclusions of the present work.

4. References

[1] S. L. Shapiro and S. A. Teukolsky, *Black Holes, White Dwarfs and Neutron Stars* (Wiley, New York, 1983); N. K. Glendenning, *Compact Stars: Nuclear Physics, Particle Physics and General Relativity*, 2nd ed.

(Springer, Berlin, 2000); P. Haensel, A. Y. Potekhin and D. G. Yakovlev, *Neutron Stars 1: Equation of State and Structure* (Astrophysics and Space Science Library, Springer, New York, 2007).

[2] V. A. Ambartsumyan and G. S. Saakyan, Sov. Astron. **4**, 187 (1960).

[3] N. K. Glendenning, Astrophys. J **293**, 470 (1985); N. K. Glendenning and S. A. Moszkowski, Phys. Rev. Lett. **67**, 2414 (1991).

[4] F. Weber and M. K. Weigel, Nucl. Phys. A **505**, 779 (1989).

[5] J. Schaffner and I. Mishustin, Phys. Rev. C **53**, 1416 (1996).

[6] S. Balberg and A. Gal, Nucl. Phys. A **625**, 435 (1997).

[7] M. Baldo, G. F. Burgio and H.-J. Schulze, Phys. Rev. C **61**, 055801 (2000).

[8] I. Vidaña, A. Polls, A. Ramos, L. Engvik and M. Hjorth-Jensen, Phys. Rev. C **62**, 035801 (2000).

[9] H.-J. Schulze, A. Polls, A. Ramos and I. Vidaña, Phys. Rev. C **73**, 058801 (2006).

[10] H. Dapo, B.-J. Schaefer and J. Wambach, Phys. Rev. C **81**, 035803 (2010).

[11] S. E. Thorsett and D. Chakrabarty, Astrophys. J. **512**, 288 (1999).

[12] T. Takatsuka *et al.,* Eur. Phys. J. A **13**, 213 (2002); T. Takatsuka *et al.,* Prog. Theor. Phys. Supp. **174**, 80 (2008).

[13] H.-J. Schulze *et al.*, Phys. Lett. B **355**, 21 (1995); Phys. Rev. C **57**, 704 (1998); M. Baldo *et al.,* Phys. Rev. C **58**, 3688 (1998); I. Vidaña *et al.,* Phys. Rev. C **61**, 025802 (2000).

[14] H. Q. Song *et al.,* Phys. Rev. Lett. **81**, 1584 (1998); Phys. Lett. B **473**, 1 (2000).

[15] R. B. Wiringa, V. G. J. Stoks and R. Schiavilla, Phys. Rev. C **51**, 38 (1995).

[16] P. M. M. Maessen, Th. A. Rijken and J. J. de Swart, Phys. Rev. C **40**, 2226 (1989).

[17] B. A. Loiseau, Y. Nogami and C. K. Ross, Nucl. Phys. A **165**, 601 (1971); Erratum A **176**, 665 (1971); P. Grangé *et al.,* Phys. Lett. B **60**, 237 (1976);

[18] X. R. Zhou *et al.,* Phys. Rev. C **69**, 018801 (2004); Z. H. Li *et al.,*Phys. Rev. C **77**, 034316 (2008); Z. H. Li and H.-J. Schulze, Phys. Rev. C **78**, 028801 (2008).

[19] J. D. Chalk III and B. W. Downs, Phys. Rev. **132**, 2727 (1963).

[20] E. M. Nyman, Phys. Rev. **159**, 782 (1967).

[21] A. Gal, Phys. Rev. Lett. **18**, 568 (1967).

[22] R. K. Bahaduri, B. A. Loiseau and Y. Nogami, Nucl. Phys. B **3**, 380 (1967); Ann. Phys. **44**, 57 (1967).

[23] B. A. Loiseau, Nucl. Phys. B **9**, 169 (1969).

[24] A. R. Bodmer, Q. N. Usmani and J. Carlson, Phys. Rev. C **29**, 684 (1984).

[25] Y. Yamamoto, Phys. Rev. C **36**, 2166 (1987).

[26] M. Kimura and W. Satoh, Prog. Theor. Phys. **88**, 605 (1992).

[27] Q. N. Usmani and A. R. Bodmer, Phys. Rev. C **60**, 055215 (1999).

[28] D. H. Youngblood, H. L. Clark, and Y.-W. Lui, Phys. Rev. Lett. **82**, 691 (1999).

[29] D. J. Millener, C. B. Dover and A. Gal, Phys. Rev. C **38**, 2700 (1988).

[30] P. Demorest *et al.,* Nature **467**, 1081 (2010).

[31] D. J. Champion *et al.,* Science **320**, 1309 (2008); P. C. C. Freire, arXiv:0907.3219; P. C. C. Freire, arXiv:1011.5809.

[32] R. A. Hulse and J. H. Taylor, Astrophys. J. Lett. **195**, L51 (1975); J. H. Taylor, Phil. Trans. R. Soc. London A **341**, 117 (1992).

[33] J. Mareš, E. Friedman, A. Gal, Prog. Theor. Phys. Suppl. **117**, 227 (1994).

[34] J. Dabrowski, Phys. Rev. C **60**, 025205 (1999).

[35] H. Noumi *et al.,* Phys. Rev. Lett. **89**, 072301 (2002); **90**, 049902 (E) (2003).

[36] P.K. Saha *et al.,* Phys. Rev. C **70**, 044613 (2004).

Journal of Physics: Conference Series **342** (2012) 012007 doi:10.1088/1742-6596/342/1/012007

Is the apparent dichotomy between bursting activity of magnetars and radio pulsars real ?

J. A. Pons[1] **and R. Perna**[2]

[1] Departament de Fisica Aplicada, Universitat d'Alacant, Ctra. San Vicent del Raspeig, s/n. 03690 San Vicent del Raspeig, Spain

[2] Department of Astrophysical and Planetary Science, University of Colorado at Boulder, 440 UCB, Boulder, CO, 80304

E-mail: jose.pons@ua.es

Abstract.
Anomalous X-ray Pulsars (AXPs) and Soft Gamma-Ray Repeaters (SGRs) are a class of young neutron stars (NSs) characterized by high X-ray quiescent luminosities, short X-ray bursts, and giant flares (for SGRs). They are believed to be magnetars, i.e. NSs with magnetic fields $\sim 10^{14} - 10^{15}$ G. The discovery of magnetar-like X-ray bursts from the young pulsar PSR J1846-0258 [1], with an inferred surface dipolar magnetic field of $B_p = 4.9 \times 10^{13}$ G, lower than the traditionally considered magnetar range, and, more recently, by the discovery of SGR 0418+5729 with an even lower $B_p = 7.5 \times 10^{12}$ G [2], well within the range of the rotation powered pulsars which do not display any bursting behaviour, has raised the obvious question: why some "high-B" pulsars (PSR J1119-6127 and PSR J1814-1744, with $B \sim 4 - 5 \times 10^{13}$ G) do not display any burst, while at least one case of "low-B" NSs (SGR 0418+5729) does, if the magnetic field is their driving force ?

1. Introduction

The majority of NSs are observed as radio-pulsars, with magnetic fields in the $\sim 10^{11} - 10^{13}$ G range. They spin down due to dipole radiation losses, while converting a fraction of their rotational energy into electromagnetic radiation. The conventional wisdom is that, when the star is not accreting, its main energy source comes from rotational energy. This was shattered by the discovery of a peculiar class of sources characterized by long periods ($P \sim 2 - 11$ s) and high quiescent X-ray luminosities ($L_x \sim 10^{33} - 10^{35}$ erg s^{-1}), generally larger than their entire reservoir of rotational energy. These sources, historically classified as AXPs and SGRs, are also characterized by stochastic bursts of X-rays, releasing energies $\sim 10^{39} - 10^{41}$ erg, although some cases do not show such activity (1E 1841-045). In the case of the SGRs, they are also characterized by more energetic γ-ray flares, with typical energetics $\sim 10^{44} - 10^{45}$ erg. Several pieces of observations have allowed to establish that these peculiar objects are isolated, and hence the extra energy cannot be provided by accretion from a companion.

The most successful model to explain both the high quiescent X-ray luminosities, as well as the X-ray bursts and giant γ-ray flares is the *magnetar* model [3, 4], in which SGRs and AXPs are believed to be endowed with large magnetic fields, $B \sim 10^{14} - 10^{15}$ G. After a magnetar is born, the internal magnetic field is subject to a continuous process of evolution through the processes of Ohmic dissipation, Ambipolar diffusion, and Hall drift. Soon after birth (from hours

Journal of Physics: Conference Series **342** (2012) 012007 doi:10.1088/1742-6596/342/1/012007

to days, at most) a solid crust is formed and protons in the liquid core undergo a transition to a superconducting phase. At this stage, the field evolution is likely to be governed by crustal processes, because the role of ambipolar diffusion in a superconducting core is questionable [5]. Dissipation of fields in the magnetar range can account for the enhanced persistent luminosities of AXPs and SGRs. These ideas have been confirmed by numerical simulations of coupled magneto-thermal evolution in the crust of NSs [6, 7].

The magnetic field also plays a fundamental role in the production of bursts and flares. The physical mechanisms responsible for these have been identified and amply discussed by [3, 4]. The remarkable analogy between tectonic activity of NSs ("starquakes") and earthquakes on Earth was soon realized, because both phenomena share some statistical properties [8]. In the crust, magnetic stresses are generally balanced by elastic stresses. However, as the internal field evolves, local magnetic stresses can become too strong to be balanced by the elastic strength of the crust, which hence breaks. Then, after the starquake, a new equilibrium state is reached and the extra stored magnetic/elastic energy becomes available for powering the bursts and flares. According to the model by [3, 4], giant flares are likely to be the result of a global event, while small bursts are local phenomena.

Despite the success of the magnetar model in explaining some general features of the triggering mechanism of bursts and flares, some major questions have been left unanswered. In particular: what determines the frequency of the bursts? Why some objects display giant flares (SGRs), while others (AXPs) do not? How frequent are the different phenomena ? Why does the burst location on the NS surface correlate with the pulsar phase (maximum of the quiescent X-ray lightcurve) for AXPs, while it is uncorrelated with phase for SGRs? (see e.g. [9] for a review]). It is clear that the magnetic field alone, or at least the dipolar field component inferred through measurements of the period P and its derivative $\dot{P}$, is not sufficient to account for this variety of behaviours that pulsars manifest. A *unified* physical framework is still lacking, and it constitutes a major puzzle in NS physics. Here we present a summary of recent works [10, 11] where we have provided the first estimates of the burst energetics and recurrence times based on long-term 2D simulations of the magneto-thermal evolution of the NS with realistic values of the breaking stress of the NS crust. Our approach allows us to identify some key elements of the magnetar phenomenology, and to shed light on what creates the variety of the – often puzzling – observed behaviour.

2. Our approach to the problem.

We follow the evolution (in axial symmetry) of the magnetic field in a magnetar crust with the numerical code described in [12]. The initial configurations of the magnetic field include both a toroidal and a poloidal component. The temperature of the crust at different ages is given by the results of [7], where the interplay between the magnetic field evolution and the thermal evolution of magnetars was studied. For simplicity, and for numerical limitations, here we assume that the crust is isothermal. As the magnetic field evolves, the elastic crust moves through a series of equilibrium states in which its elastic stress $\sigma_b(r, \theta, t)$ balances the (time-dependent) magnetic stress $M_{ij}(r, \theta, t)$ in each direction. Assuming that equilibrium is reached at a certain time, $t^{\rm eq}$, we can define

$$M_{ij}^{\rm eq}(r,\theta) = \frac{B_i(r,\theta,t^{\rm eq})B_j(r,\theta,t^{\rm eq})}{4\pi} \sim \sigma_b(r,\theta,t^{\rm eq}) \, , \qquad (1)$$

where r and θ indicate the radial and poloidal coordinates, respectively.

Recent molecular dynamical simulations [13] have provided a fit for the maximum stress that a NS crust can sustain

$$\sigma_b^{\rm max} = \left(0.0195 - \frac{1.27}{\Gamma - 71}\right) n_i \, \frac{Z^2 e^2}{a} \, , \qquad (2)$$

where $\Gamma = Z^2e^2/aT$ is the Coulomb coupling parameter, $a = [3/(4\pi n_i)]^{1/3}$ is the ion sphere radius, n_i the ion number density, Z the charge number, e the electron charge, and T is the temperature in energy units. This fit to results from extensive molecular dynamics was found to be in agreement with the Zhurkov model of strength, and provides a good approximation to the breaking stress for very long timescales ≈ 1 s -1 yr. We will extrapolate this result to even longer times, of sometimes > 100 yr, but in any case, the breaking stress at very long timescales cannot exceed our adopted values [13].

As the magnetic field evolves under the influence of the dominant Hall drift term, and by Ohmic diffusion, local magnetic stresses in the crust can occasionally, at, say, a time t_b, depart from the equilibrium condition at a previous time $t^{\rm eq}$ by an amount which is comparable to, or exceeds, the breaking stress of the crust,

$$M_{ij}(r,\theta,t_b) - M_{ij}^{\rm eq}(r,\theta) \sim \sigma_b^{\rm max}(r,t_b)\ , \tag{3}$$

and hence the crust fractures. Here $\sigma_b^{\rm max}$ is independent of the angle because of our assumption of isothermality,

Since our simulations follow the evolution of $\mathbf{B}(r,\theta,t)$, and hence $M_{i,j}(r,\theta,t)$, we can map the time-dependent regions in the NS crust which are subject to fractures. When this condition is fulfilled, a starquake will happen and a new equilibrium state will be reestablished. A fully consistent dynamical simulation of the NS starquake is out of our present capabilities. The typical timescales (msec to sec) of individual burst/flare events are many orders of magnitude smaller than the long-term evolution timescales (the typical timesteps in our code are a few days or a week, to be able to follow the NS evolution up to $10^5 - 10^6$ years), hence we cannot follow dynamically the fracture propagation and model individual bursts. However, we can estimate the energy of an "outburst" (i.e. a collection of tens to hundreds of bursts occurring within $\sim$ a week timescale) as follows. When a starquake happens, say at time $t = t_b$, a certain region of the crust in the vicinity of the point where critical conditions are reached will be affected. We consider that all surrounding regions where local magnetic stresses are close to the maximum (say by a fraction $\epsilon \sim 90\%$ or 95%) will be affected, i.e.

$$M_{ij}(r,\theta,t_b) - M_{ij}^{\rm eq}(r,\theta) > \epsilon\sigma_b^{\rm max}(r,t_b)\ . \tag{4}$$

We can compute the elastic energy stored in that portion of the crust volume, and assume that it will be released at once (i.e., in one timestep, ≈ 1 week) and that the affected region will return immediately to equilibrium. The new equilibrium stresses are reset, and the process is repeated as the magnetic field evolves (we neglect the feedback produced by the local deposition of energy, which will require a much more complex modeling). This allows us to obtain the energy available in a given event, the time interval between events, and the location of the fracture. It should be stressed that this represents the total energy of an "outburst", that may be released in one very energetic flare, or many small bursts, or a combination of both. Also, this estimate is to be taken as an upper limit to the event energetics. Depending on the location and local physical conditions (density, temperature), part of the energy is lost by neutrino emission, part of it can be transferred to the magnetosphere, and part of it results in local heating and radiated by photons from the surface on a much longer timescale. Typically, energy released in the inner crust (above neutron drip) is more easily lost in the form of neutrinos while energy released near the surface can have a very high efficiency and a direct observational impact in thermal (surface) and non-thermal (magnetospheric) radiation.

The energy available after a starquake can therefore be estimated as the elastic energy corresponding to a static shear strain Ψ, which can be well approximated by [4]

$$E_b(t_b) = \frac{1}{2}\int dV\mu\Psi^2 = \Sigma_{i,j}\int dV\frac{\left[M_{ij}(r,\theta,t_b) - M_{ij}^{\rm eq}(r,\theta)\right]^2}{\mu(r,t_b)} \tag{5}$$

where we have used the fact that the yield strain of the crust in the region where breaking conditions are fulfilled is approximated by

$$\Psi_{ij} \approx \frac{\left[M_{ij}(r,\theta,t_b) - M_{ij}^{\rm eq}(r,\theta)\right]}{\mu(r,t_b)}, \tag{6}$$

and the shear modulus μ is given by [14]

$$\mu(r,t) = \left(0.01106 - \frac{28.7}{\Gamma^{1.3}}\right) n_i \frac{Z^2 e^2}{a}. \tag{7}$$

The integral in Eq.(5) is computed over the volume dV for which condition (4) on the stresses in any direction is satisfied.

3. Results.

We now report the most relevant results of a number of cases that we studied. As a baseline model we have chosen a NS born with an initial dipolar field of $B_p = 8 \times 10^{14}$ G (at the pole) and an internal toroidal field of $B_t = 2 \times 10^{15}$ G at maximum, similar to models used in previous studies and close to what one expects to be the final geometry after MHD equilibrium is reached in a hot, liquid proto-neutron star. In this particular model, at the age of 10^4 yrs the dipolar field has decreased to about 20% of its initial value, while its internal toroidal field has been dissipated by more than a factor of 2. Of course other initial conditions modify quantitatively the result we show in the following, but the general trends remain nearly the same. The evolution is followed for 10^5 years. During this time, we monitor the frequency and angular distribution of magnetic stresses in the NS crust, and the energy released any time local magnetic stresses become strong enough to fracture the crust.

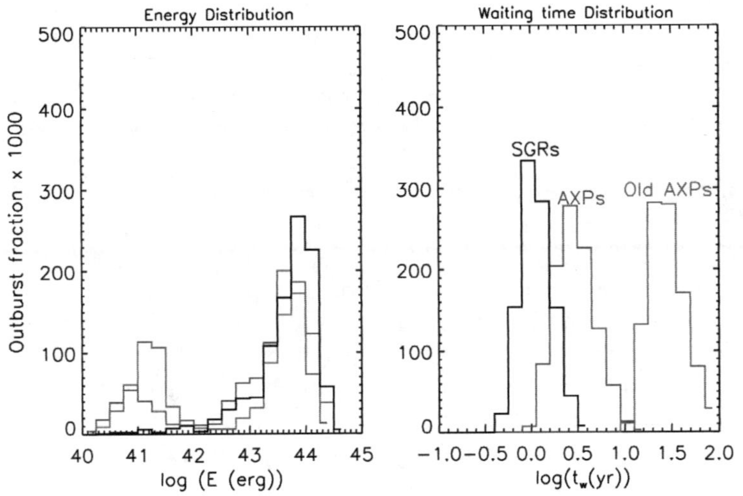

Figure 1. Outburst properties of an object with initial magnetic field components $B_p = 8 \times 10^{14}$ G and $B_t = 2 \times 10^{15}$ G during three different periods of its lifetime: 400-1600 yr (labeled "SGR"), 7-10 kyr (labeled as "AXP"), and 60-100 kry (labeled as "old AXP"). We have fixed $\epsilon = 0.9$.

The main results about frequency and distribution of energies available to be released following crust fractures are summarized in Fig. 1. We remind again that the time step of our simulation is ≈ 1 week, thus groups of bursts that would occur on timescales shorter than this are classified as a single outburst in our simulations. We selected three representative periods in a magnetar life, involving each one the same total number of events (1000). The first one is labeled "SGR" and spans the interval between 400 and 1600 yrs. The second case is labeled "AXP" and covers the period $\approx 7 - 10$ kyrs, and the third period is named "old AXP" and corresponds to the events recorded from 60 to 100 kyrs. The first important result is that there is a strong difference in the energetics and recurrence time as the star evolves. This is because the crustal magnetic field evolves more rapidly initially due to the Hall drift, while at late times its geometry has rearranged into a more quasi-steady state and the evolution is slower. In young SGRs, the typical energies released after each starquake are of the order of 10^{44} erg and the typical event rate ~ 1/yr. Note that this does not necessarily imply that a flare will be observed. We remind that this is the maximum energy available released in the interior, but the mechanism to transfer this energy to the surface and magnetosphere may not work in many cases, especially when the fracture occurs in the inner crust. From our results, we observed that more than 90% of the fractures occurs close to the surface, simply because the crust is less strong there, so that we expect a relatively high efficiency of the process. A more detailed study of the efficiency of the energy transfer is out of the scope of this work, which is aimed to describe the expected behavior in statistical terms. For "AXP-like" objects the energetics is shifted to lower values and a second peak appears at about 10^{41}ergs, and the waiting time between outbursts increases to a few years. This trend is more pronounced as the object gets older, and for old AXPs the recurrence time becomes of the order of tens of years and nearly half the events are low-energy events. We find that, independently of age, the longest waiting times correspond to the most energetic, flare-type, events ($E > 10^{44}$ erg). Hence giant flares are expected to be less frequent than the less energetic bursts. Indeed, while giant flares have been observed in only 3 objects, bursts have been observed in more than a dozen. The physical reason behind the bimodal distribution is the directionality of stresses. We found that fractures associated to the magnetic stress component $M_{\theta,\phi}$ are more frequent, but they are mostly associated to low energy events $\approx 10^{41}$ erg. On the other hand, fractures caused by the $M_{r,\phi}$ component are responsible for most of the $E > 10^{44}$ erg events. The events associated to large values of $M_{r,\theta}$ span the whole range of energies, but they become very rare after the "SGR" stage, so that the long term energy distribution becomes more clearly bimodal.

It is interesting to compare the general trends predicted by our simulations with the observations of magnetars so far. We find that, for a given B-field strength and configuration at birth, younger objects are generally more active, both in event frequency and strength, and hence a natural sequence for a magnetar is to evolve from an SGR to an AXP-like object. Interestingly, this age sequence is also what observations have been hinting at [15]. Overall, we find that the energetics of the "SGR-like" objects tend to be higher than the ones of the "AXP-like" objects, again in agreement with observations [16]. Our predicted clustering of the energy distribution (around $10^{40} - 10^{41}$ and 10^{44} erg) is suggestive of the observed dichotomy outbursts/giant flares, with much rarer events in between [17]. Our waiting time distributions are roughly log-normal. A direct comparison with observations cannot however be made at this stage, due to the very limited statistics. However, if the secular waiting time distribution of starquakes (which is what we predict) has a similar behaviour to the waiting time distribution of bursts within an outburst, then our results would again match the observations [8].

Further connections to the observations can be made by inspection of the right panels of Fig. 1, where we show the angular distribution of the bursts. While for "SGRs" and "old AXPs" there is no clear trend, it is interesting to note that for the sample named "AXPs" most of the fractures happen at small polar angles. If the polar region is hotter than the equator, our results suggest a correlation between burst location and pulsar phase. The properties of this object are typical of what is routinely classified as an AXP [9]. The reason for this particular feature at middle ages is probably related to the evolution of the internal toroidal field, as already discussed in [12]. The Hall drift displaces the internal toroidal field towards the poles, in a typical timescale of $10^3 - 10^4$ yrs (the Hall timescale), but after a few Hall timescales the magnetic field is reconfigured in a more stable, steady-state. Maybe this characteristic location of starquakes closer to the pole is reflecting this fact, but we cannot make a definite claim about this point. Consistent simulations that include the local heating and the temperature variations produced by the energy release are in progress and will be presented elsewhere.

4. Conclusions.

The results from the study cases presented in this letter bear a direct relevance for the interpretation of the overall magnetar phenomenology. Overall, an "SGR-like" object tends to have a higher dipolar field than an "AXP-like" object or a "high-B radio pulsar", but there is no fundamental separation among what constitutes the apparent different classes. Among the key elements that create the variety of observed phenomena, the age is probably more important that a small variation in the magnetic field strength. SGR-like flares can be produced even with low B_p if the internal toroidal field is sufficiently large, but those events are in any case much less frequent in old, low-field NSs than in young, high-field magnetars.

In summary, we have found that there is no fundamental separation among what constitutes the apparently different classes of "normal radio pulsars", of "high-B radio pulsars", of "high-B AXPs', "high-B SGRs", or "low-B SGRs/AXPs". We have been able to identify some key elements that create the variety of observed phenomenology. In particular, while what we infer from measurements of P and $\dot{P}$ is only the dipolar B-field component, B_p, both the dipolar component and the strength of the hidden toroidal component are similarly important. The importance of a toroidal component had already been pointed out in the earlier work of [4]; they noticed how an internally wound B-field can more easily stress the NS crust, as our results have shown, and it is also a natural outcome of a strong α-dynamo. More importantly, our results show that the age (or the evolutionary stage) is the most relevant feature to understand different levels of activity in different families of NSs. The same object may behave differently at different times, without in principle having a different nature. Our simulations have been able to quantify this effect, explain key observational properties, and give a step forward to connect and unify the entire pulsar/magnetar phenomenology.

References

[1] Gavriil F P, Rim D, and Kaspi V M 2008 *Science* **319** 1802
[2] Rea N et al. 2010 *Science*, **330** 944
[3] Thompson C and Duncan R C 1995 *MNRAS* **275** 255
[4] Thompson C and Duncan R C 2001 *ApJ* **561** 980
[5] Glampedakis K, Jones D I and Samuelsson L 2010 *MNRAS* **413** 2021
[6] Aguilera D N, Pons J A and Miralles J A 2008 *ApJ* **673**, 167
[7] Pons J A, Miralles J A and Geppert U 2009 *A&A* **496** 207
[8] Cheng B, Epstein R I, Guyer R A and Cody Young A 1995 *Nature* **382** 518
[9] Kaspi V M and Gavriil F P 2004 *Nucl. Phys. B* **132** 456
[10] Perna R and Pons J A 2011 *ApJ* **727** L51
[11] Pons J A and Perna R 2011 *ApJ* **741** 123
[12] Pons J A and Geppert U 2007 *A&A* **470** 303
[13] Chugunov A I and Horowitz C J 2010 *MNRAS* **407** L54
[14] Horowitz C J and Hughto J C, eprint arXiv: 0812.2650.
[15] Kouveliotou C 1999 *Proc. Nat. Academy of Science* **96**, 5351
[16] Gavriil F P, Kaspi V M, and Woods P M 2004 *ApJ* **607** 959
[17] Kouveliotou C et al. 2001 *ApJ Lett.* **558** L47

2nd Iberian Nuclear Astrophysics Meeting on Compact Stars | IOP Publishing
Journal of Physics: Conference Series **342** (2012) 012008 | doi:10.1088/1742-6596/342/1/012008

Effect of the symmetry energy on compact stars

Constana Providência[1], Rafael Cavagnoli[1,2], Debora P. Menezes[2], Prafulla K. Panda[1,3]

[1]Centro de Física Computacional, Department of Physics, University of Coimbra, P3004-516 Coimbra, Portugal

[2]Departamento de Física, CFM,Universidade Federal de Santa Catarina, Florianópolis - SC - CP. 476, CEP 88.040 - 900, Brazil

[3]Department of Physics, C.V. Raman College of Engneering, Vidya Nagar, Bhubaneswar-752054, India

E-mail: cp@fis.uc.pt

Abstract. The effect of the symmetry energy on the properties of compact stars is discussed. It is shown that, for stars with masses above 1 $M_\odot$, the radius of the star varies linearly with the symmetry energy slope L. The hyperon-meson couplings are chosen according to experimental values of the hyperon nuclear matter potentials, and possible uncertainties are considered. It is shown that a softer symmetry energy gives rise to stars with less hyperons and smaller radius. Hyperon-meson couplings may also have a strong effect on the mass of the star.

1. Introduction

In the present work we want to understand how sensitive is the mass/radius curve of a family of stars to the symmetry energy and its slope at saturation. We study not only maximum mass configurations but also stars with a mass in the range $1.0\,M_\odot < M < 1.4\,M_\odot$. These stars have a central density that goes from 1.5 ρ_0 to 2-3 ρ_0, and, therefore, we will be testing the equation of state at suprasaturation densities. At high density the formation of hyperons is energetically favorable and, therefore, we also study the effect of the symmetry energy on the appearance of these exotic degrees of freedom [1].

The work is performed in the relativistic mean field approximation and we consider different parametrizations of the non-linear Walecka model [2], which allow us to discuss the role of the density dependence of the symmetry energy on the properties of the non-homogeneous nuclear EOS [1, 3, 4, 5, 6]. We will also present results obtained with the quark-meson-coupling (QMC) model [7, 8], an effective nuclear model that takes into account the internal structure of the nucleon explicitly. Within the QMC model, matter at low densities and temperatures is a system of nucleons interacting through meson fields, with quarks and gluons confined within MIT bags.

We consider three different hyperon-meson parametrizations: one proposed in [1] and a second one that takes into account the different binding energies of the hyperons [10, 11], and the third one starts from the QMC model and also takes into account the different binding energies of the hyperons. The results presented here have been published in part in [12] and [13].

2. Formalism

We describe hadronic matter within the framework of the relativistic non-linear Walecka model (NLWM) [2]. We include a $\rho - \omega$ meson coupling term as in [5, 6] in order to study the effect of the symmetry energy on the star properties while leaving the isoscalar channel fixed. The Lagrangian density reads

$$\begin{aligned}\mathcal{L} &= \sum_{j=1}^{8} \bar{\psi}_j \left[\gamma_\mu \left(i\partial^\mu - g_{\omega j}\, \omega^\mu - g_{\rho j}\, \vec{\tau}_j \,.\, \vec{\rho}^{\,\mu}\right) - m_j^* \right] \psi_j + \sum_{l=1}^{2} \bar{\psi}_l \left(i\gamma_\mu \partial^\mu - M_l\right) \psi_l \\ &+ \frac{1}{2}\partial_\mu \sigma \partial^\mu \sigma - \frac{1}{2} m_\sigma^2 \sigma^2 - \frac{1}{3!} k \sigma^3 - \frac{1}{4!}\lambda \sigma^4 - \frac{1}{4}\Omega_{\mu\nu}\,\Omega^{\mu\nu} + \frac{1}{2} m_\omega^2\, \omega_\mu \omega^\mu + \frac{1}{4!}\xi g_\omega^4 (\omega_\mu \omega^\mu)^2 \\ &- \frac{1}{4}\vec{R}_{\mu\nu} \,.\, \vec{R}^{\mu\nu} + \frac{1}{2} m_\rho^2\, \vec{\rho}_\mu \,.\, \vec{\rho}^{\,\mu} + \Lambda_{\rm v} (g_\rho^2\, \vec{\rho}_\mu \,.\, \vec{\rho}^{\,\mu})(g_\omega^2\, \omega_\mu \omega^\mu)\,, \end{aligned} \tag{1}$$

$\Omega_{\mu\nu} = \partial_\mu \omega_\nu - \partial_\nu \omega_\mu$, $\vec{R}_{\mu\nu} = \partial_\mu \vec{\rho}_\nu - \partial_\nu \vec{\rho}_\mu - g_\rho\, (\vec{\rho}_\mu \times \vec{\rho}_\nu)$, g_{ij} are the coupling constants of mesons $i = \sigma, \omega, \rho$ with baryon j, m_i is the mass of meson i and l represents the leptons e^- and μ^-. The couplings k and λ are the weights of the non-linear scalar terms and $\vec{\tau}$ is the isospin operator. Their values can be obtained from [12]. The sum over j in (1) extends over the octet of lightest baryons $\{n, p, \Lambda, \Sigma^-, \Sigma^0, \Sigma^+, \Xi^-, \Xi^0\}$.

In the QMC model, the nucleon in nuclear medium is assumed to be a static spherical MIT bag in which quarks interact with the scalar (σ) and vector (ω, ρ) fields, and those are treated as classical fields in the mean field approximation (MFA) [7, 8, 9]. The quark field, ψ_{q_i}, inside the bag then satisfies the equation of motion:

$$\left[i\,\partial\!\!\!/ \; - \; (m_q^0 - g_\sigma^q\,) - g_\omega^q\, \omega\, \gamma^0 + \frac{1}{2} g_\rho^q \tau_z \rho_{03} \gamma^0\right] \psi_{q_i}(x) = 0\,, \quad q = u, d, s \tag{2}$$

where m_q^0 is the current quark mass, and g_σ^q, g_ω^q and g_ρ^q denote the quark-meson coupling constants. The effective mass of a nucleon bag at rest is taken to be equal to the energy of the bag, $M_i^* = E_i^{\rm bag}$. The equilibrium condition for the bag is obtained by minimizing the effective mass, M_i^* of baryon i with respect to the bag radius, see [7, 8, 9].

$$\frac{d\,M_i^*}{d\,R_i^*} = 0, \tag{3}$$

The energy density and pressure of hadronic matter within QMC is equal to the one obtained for NLWM with the effective mass replaced by the energy of the bag, $E_i^{\rm bag}$. We will also include a meson nonlinear term involving the ω and ρ mesons. Choosing the coupling constant Λ_v adequately it is possible to soften the symmetry energy of the QMC model at large densities.

We consider two different sets of hyperon-meson couplings. For the coupling set A the ω and ρ meson-hyperon coupling constants are obtained using the SU(6) symmetry: $\frac{1}{2} g_{\omega\Lambda} = \frac{1}{2} g_{\omega\Sigma} = g_{\omega\Xi} = \frac{1}{3} g_{\omega N}$, $\frac{1}{2} g_{\rho\Sigma} = g_{\rho\Xi} = g_{\rho N}$, $g_{\rho\Lambda} = 0$, where N means 'nucleon' ($g_{iN} \equiv g_i$). The coupling constants $\{g_{\sigma j}\}_{j=\Lambda,\Sigma,\Xi}$ of the hyperons with the scalar meson σ are constrained by the hypernuclear potentials in nuclear matter to be consistent with hypernuclear data [10]. We take $V_j = x_{\omega j}\, V_\omega - x_{\sigma j}\, V_\sigma$ where $x_{ij} \equiv g_{ij}/g_i$, $V_\omega \equiv g_\omega \omega_0$ and $V_\sigma \equiv g_\sigma \sigma_0$ are the nuclear potentials for symmetric nuclear matter at saturation with (see Ref. [10]), $V_\Lambda = -28$ MeV, $V_\Sigma = 30$ MeV , $V_\Xi = -18$ MeV. We use these same values in QMC. In QMC the couplings of the hyperons to the σ-meson do not need to be fixed because the effective masses of the hyperons are determined selfconsistently at the bag level. We obtain $x_{\omega B}$ from the hyperon potentials in nuclear matter, $V_j = -(M_j^* - M_j) + x_{\omega j} g_\omega \omega_0$, and find $x_{\omega\Lambda} = 0.743$, $x_{\omega\Sigma} = 1.04$ and $x_{\omega\Xi} = 0.346$. $x_{\rho B} = 1$ is fixed for all the baryons [13]. However, while the binding of the Λ to symmetric

Journal of Physics: Conference Series **342** (2012) 012008 doi:10.1088/1742-6596/342/1/012008

nuclear matter is well settled experimentally, the binding values of the Σ^- and Ξ^- still have a lot of uncertainties [14]. We, therefore, test the effect of the coupling to the cascade and show results also for $V_\Xi = -10$ and 0 MeV.

In order to show how results are sensitive to the hyperon couplings we consider set B defined as proposed in [1] with x_σ=0.8 and equal for all the hyperons. We obtain the fraction x_ω with $V_j = V_\Lambda = -28$ MeV, and take the same value for all the hyperons. For the hyperon-ρ-meson coupling we consider $x_\rho = x_\sigma$.

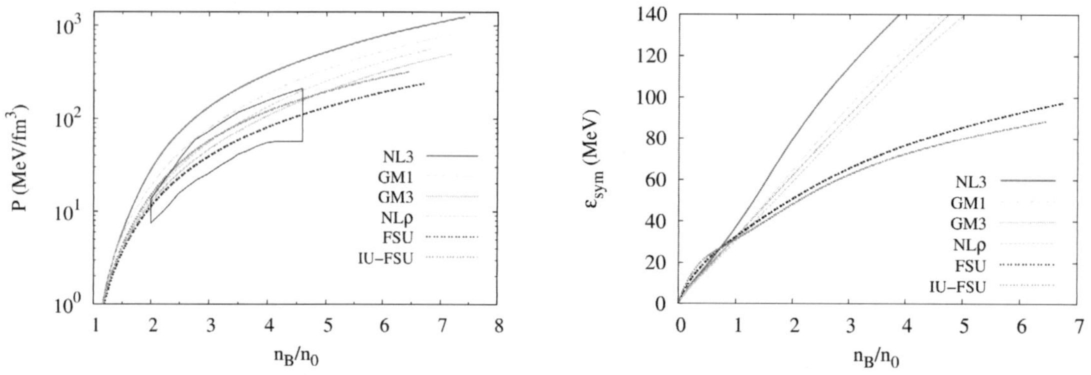

Figure 1. Left panel: Equation of state for symmetric nuclear matter and different NLWM models. The enclosed area represents experimental data according to Danielewicz *et al.*, [15]. Right panel: the corresponding symmetry energy

3. Results

Within the NLWM, we will consider the following parametrizations, see Fig. 1: NL3 [3], with a quite large symmetry energy and incompressibility at saturation and which was fitted in order to reproduce the ground state properties of both stable and unstable nuclei; FSU [5] and IUFSU [6], which were accurately calibrated to simultaneously describe the GMR in ^{90}Zr and ^{208}Pb, and the IVGDR in ^{208}Pb and still reproduce ground-state observables of stable and unstable nuclei. FSU is very soft at high densities. GM1 and GM3 [1] are generally used to describe stellar matter, with a symmetry energy not so hard as the one of NL3, and NLρ [4], which has at high densities, a behavior between GM1 and GM3.

In order to study the effect of the isovector channel in the star properties we also consider a modified version of the IUFSU parametrization: we keep the isoscalar channel and change g_ρ and Λ_v keeping the symmetry energy fixed at the density 0.12 fm^{-3}. We generate a set of models, from set 1 to set 7, that differ in their value of the symmetry energy and corresponding slope at saturation but have the same isoscalar properties. The set 1 is the parametrization IUFSU, $L = 47$ MeV. The other parametrizations have a larger symmetry energy and a larger slope L at saturation. The maximum value of L we have considered (for set 7, $L = 99$ MeV) is within the experimental values obtained from isospin diffusion in heavy ion reactions [16] and corresponds to a symmetry energy slightly softer than NLρ (see Fig. 1). Values of L below 47 MeV would give unacceptable EOS because they would predict that neutron matter is bound.

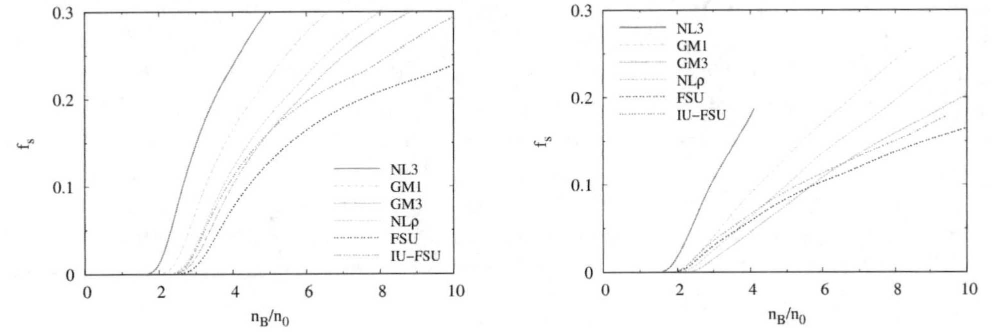

Figure 2. Strangeness fraction for different NLWM models with hyperon set A (left panel) and set B (right panel).

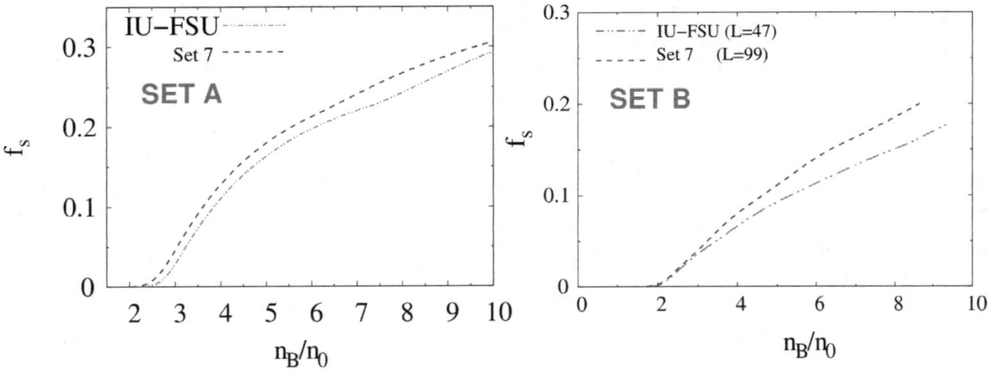

Figure 3. Strangeness content for the modified IUFSU models, with set A and set B for the hyperon-meson couplings.

The effect of the symmetry energy on the strangeness fraction is seen in Figs. 2: the strangeness content is sensitive to the model and the meson-hyperon couplings. In general, the softer the EOS the larger the strangeness onset density and the smaller the strangeness content. A large meson-hyperon vector coupling, as occurs in set B, hinders the formation of hyperons. However, we should point out that when comparing different models we are comparing not only the effect of the density dependence of the symmetry energy but also the density dependence of the isoscalar channel, namely the incompressibility and effective mass. This explains why FSU, a softer EOS, has a smaller strangeness content than IUFSU, a model with a softer symmetry energy than FSU above saturation density. In Fig. 3 we show the strangeness content for the modified IUFSU models, using set A and set B for the hyperon couplings. We conclude that the smaller the symmetry energy the smaller the hyperon content.

Cooling of the star by neutrino emission can occur relatively fast if the direct Urca process, $n \to p + e^- + \bar{\nu}_e$, is allowed [17]. The direct Urca (DU) process takes place when the proton fraction exceeds a critical value $x_{\rm DU}$[17]. Cooling rates of neutron stars seem to indicate that this fast cooling process does not occur and, therefore, a constraint is set imposing that the direct Urca process is only allowed in stars with a mass larger than $1.5\,M_\odot$, or a less restrictive limit, $1.35\,M_\odot$ [18]. Since the onset of the direct Urca process is closely related with the density dependence of the symmetry energy, this constraint gives information on the isovector channel of the EOS.

The effect of the symmetry energy and the hyperon content on the onset density of the

Journal of Physics: Conference Series **342** (2012) 012008 doi:10.1088/1742-6596/342/1/012008

nucleon direct Urca process is seen in figures 4 as function of the slope L for the IUFSU and modified versions in the left panel , and for NL3, GM1, GM3, NLρ, FSU and IUFSU models in the right panel. We conclude that: a) for matter without hyperons the larger the L the smaller the neutron-proton asymmetry above the saturation density and, therefore, the smaller the direct Urca onset density; b) the larger the slope the smaller the onset density because a larger L corresponds to a harder symmetry energy and, therefore, larger fractions of protons are favored; c) generally, for a low value of L the presence of hyperons decreases the onset density. However, the effect of the inclusion of hyperons depends on the hyperon-meson coupling.

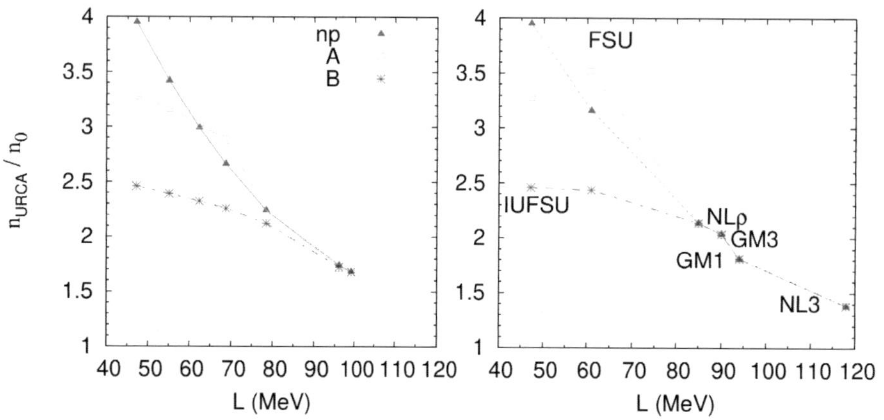

Figure 4. Onset of the direct Urca process in stellar matter for the modified IUFSU models (left panel) and NLWM (right panel), for nohyperon matter (red triangles), hyperon coupling set A (green squares) and hyperon coupling set B (blue stars).

Integrating the Tolman-Volkoff-Oppenheimer [20] equations we have obtianed the mass-radius curves for the families of stars from the EOS without hyperons and from the EOS with the meson-hyperon sets A and B.

All the models except FSU are able to describe the mass of pulsar J1614-2230 [21] if hyperons are not included [12]. However, only IUFSU and FSU satisfy the constraints set by the an empirical EOS obtained from a heterogeneous set of seven neutron stars with well-determined distances [19]. The properties of the set of relativistic mean field (RMF) models chosen are reflected on the star properties: the harder models like NL3 and GM1 predict larger masses and radius, the softest EOS, FSU, the smallest mass, the smaller the L the smaller the radius [12]. This last property is clearly seen in the left panel of Fig. 5 were the radius of maximum mass stars (squares), 1.4 $M_\odot$ stars (circles) and $1.0M_\odot$ stars (triangles) are plotted as a function of the symmetry energy slope for nucleonic stars. The full symbols are for the modified IUFSU models and the empty ones for the NL3, GM1, GM3, NLρ, FSU models. The modified IUFSU models show that if the isoscalar channel is left unchanged the radius decreases if L decreases. This reduction is larger for $1.0M_\odot$ stars (more than 1 km for $45 < L < 100$ MeV) but even for the maximum mass configurations there is still a 0.5 km difference. The set of RMF models chosen also show the same trend. However, since the isoscalar properties differ among the models, and they also affect the radius, the linear behavior is not present.

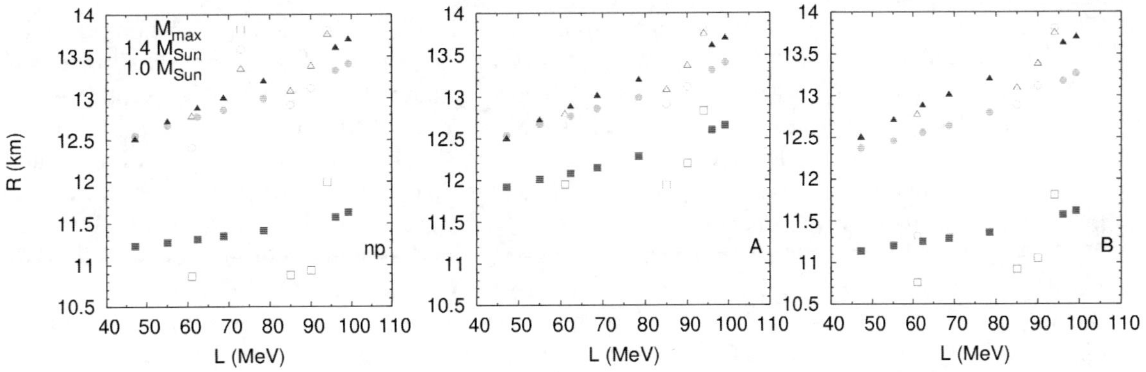

Figure 5. Star radius dependence on the slope of the symmetry energy for stars with maximum mass (squares), 1.4 $M_\odot$ (circles), 1 $M_\odot$ (triangles): no-hyperons (left panel) with hyperons, coupling A (middle panel) with hyperons, coupling B (right panel). The empty symbols are for the set of different models considered with hyperon coupling A. The full symbols are for the modified IUFSU models.

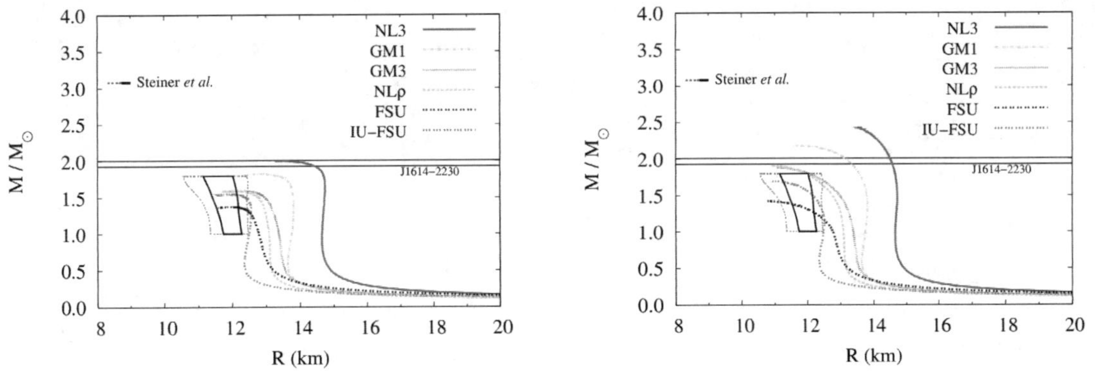

Figure 6. Mass/radius curves using set A (left panel) and set B (right panel) for the hyperon couplings..

Including hyperons in the EOS makes the EOS softer at large densities and the mass of the maximum mass stars is smaller [22]. This is seen in Fig. 6 were the mass-radius curves obtained with hyperon-meson coupling set A and B are plotted, respectively, in the left and right panels. We conclude that it is important to have correct couplings since the star masses are sensitive to the hyperon couplings. For set A only the NL3 model is able to describe the PSR J1316-2230 a star with a mass 1.97 ± 0.4 $M_\odot$ [21], while within set B NL3, GM1, GM3 and NLρ are able to describe this star. The trend discussed above between the star radius and L is still present in these stars, see the empty symbols in the middle and right panels of Fig. 5. This trend is confirmed by the modified IUFSU models, full symbols in the middle and right panels of Fig. 5. Stars with 1.0 and 1.4 $M_\odot$ contain no hyperons, or only a small fraction and therefore their radius do not different from the results obtained for np matter. Maximum mass stars, however, do have hyperons and their radius depends on the hyperon couplings chosen: for set A radius is larger and the maximum mass is smaller than the corresponding quantities predicted by set B. None of the modified IUFSU models with hyperons is able to describe the PSR J1614-2230. We also conclude that the mass of the maximum mass configuration is quite insensitive to the symmetry energy slope [12].

The EOS obtained with the QMC model are displayed in Fig. 7. We have considered QMC with nucleons and hyperons [9] and different values of the coupling parameter Λ_v and QMC with $\Lambda_v = 0$ with protons and neutrons only. We also include the empirical EOS obtained by Steiner *et al.* [19]. These results however, should still be considered with care because there are many uncertainties involved. The agreement of the calculated EOS with the empirical one when hyperons are included is defined by the hyperon-meson interaction and the Λ_v coupling, or, equivalently, by the symmetry energy. The QMC *pn* EOS agrees with the constraints. However, the inclusion of hyperons with the hyperon couplings obtained for the hyperon nuclear potentials taking $V_\Lambda = -28$ MeV, $V_\Sigma = 30$ MeV and $V_\Xi = -18$ MeV makes the EOS too soft. Increasing Λ_v makes the EOS harder bringing the EOS closer to the constraints defined by the empirical EOS. Increasing Λ_v corresponds to decreasing the symmetry energy at supersaturation densites, giving rise to a softer *pn* EOS at high densities and, therefore, hinders the onset of hyperons. So the larger Λ_v the smaller the hyperon fraction in the star and the harder the EOS.

The effect of a less attractive V_Ξ potential is also clear: the EOS becomes harder because the onset of hyperons occurs at larger densities. We conclude that any mechanism that hinders the formation of hyperons makes the EOS harder.

The mass/radius curve for stars with a mass larger than $1\,M_\odot$ and the corresponding properties of maximum mass stars are shown in Fig. 8.

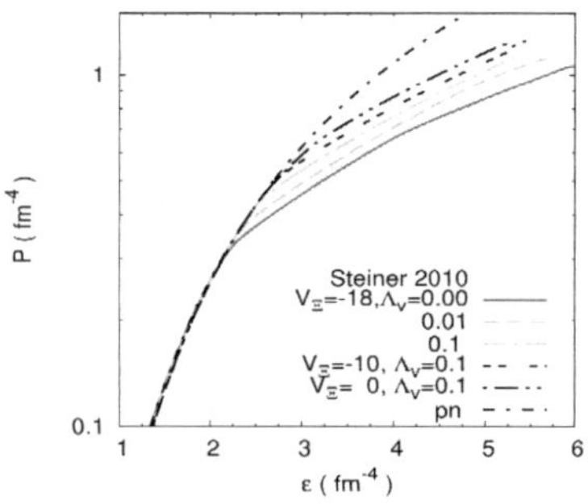

Figure 7. Equation of state for QMC and different values of V_Ξ and of the Λ_v coupling.

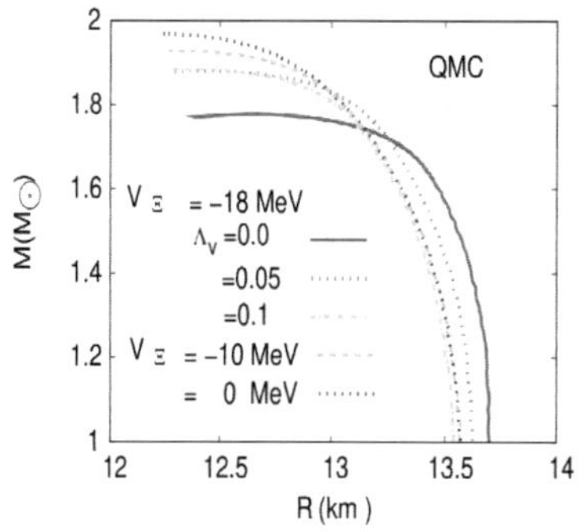

Figure 8. Mass/radius curve for the families of compact stars obtained with the EOS of Fig. 7.

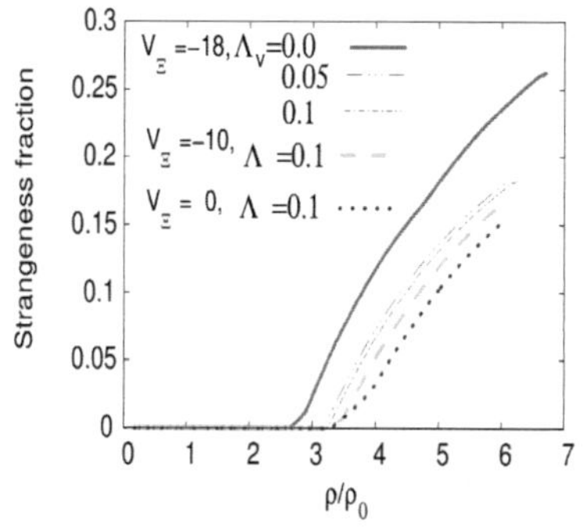

Figure 9. Strangeness content in the models represented in Fig. 7.

First let us discuss the effect of the symmetry energy and the hyperon couplings on the mass/radius curve. A larger Λ_v gives rise to a softer EOS and, therefore, a smaller radius. It is seen that when going from $\Lambda_v = 0$ to 0.1 the radius of stars with a mass $M = 1 - 1.5\,M_\odot$ decreases by ~ 0.3 Km. A similar effect was obtained with the NLWM [12, 13]. However, within the NLWM models the maximum mass did not depend on L, while in the framework of the QMC model there is a clear effect of almost $0.1 M_\odot$ if Λ_v increases from 0 to 0.1. This is mainly due to the smaller strangeness fraction inside the star.

The reduction of the attractiveness of V_Ξ has a similar effect on the maximum mass of the star, i.e., the mass increases $\sim 0.2\,M_\odot$ if V_Ξ increases from -18 to 0 MeV.

We conclude that there is still quite a large uncertainty on the coupling of hyperons to nuclear matter and, therefore, there is still room for a very massive star such as the recently measured pulsar J1614-2230 with a mass $M = 1.97 \pm 0.04$ [21], even including hyperons in the EOS. This, however, is a particularly massive star.

4. Conclusions

We have tested three different hyperon-meson parametrizations, using information from hypernuclei to fix the couplings. We took advantage of the fact that QMC predicts the hyperon effective masses without being necessary to fix the hyperon-σ couplings, in one of the parametrizations.

It was shown that both the symmetry energy and the hyperon couplings have a strong effect on the mass and radius of the star. A softer symmetry energy gives rise to smaller stars. Also the hyperon fraction is affected: softer symmetry energy corresponds to a smaller hyperon fraction [12, 13]. Within QMC the density dependence of the symmetry energy has also an effect on the maximum star mass, effect not observed in NLWM models.

It was also shown that the hyperon nuclear interaction defines the amount of strangeness in the star, and, therefore, has a strong influence on the maximum mass allowed. Even including hyperons in the QMC EOS we could explain the mass of the pulsar J1614-2230 if the cascade nuclear potential is set to be very little attractive. More data on hypernuclei is needed to constrain the hyperon-meson couplings.

Acknowledgments

Acknowledgments This work was partially supported by FCT (Portugal) under the grants PTDC/FIS/113292/2009 and CERN/FP/116366/2010 and by CNPq (Brazil) and CAPES(Brazil)/FCT(Portugal) under project 232/09.

References

[1] N. K. Glendenning and S. A. Moszkowski, Phys. Rev. Lett. **67**, 2414 (1991)
[2] B.D. Serot and J.D. Walecka, Adv. Nucl. Phys. **16** (1986) 1.; J. Boguta and A. R. Bodmer, Nucl. Phys. **A 292**, 413 (1977).
[3] G. A. Lalazissis, J. König, P. Ring, Phys. Rev. C **55**, 540 (1997).
[4] B. Liu, V. Greco, V. Baran, M. Colonna, and M. Di Toro, Phys. Rev. C **65**, 045201 (2002).
[5] B. G. Todd-Rutel and J. Piekarewicz, Phys. Rev. Lett. **95**, 122501 (2005); F. J. Fattoyev and J. Piekarewicz, Phys. Rev. C **82**, 025805 (2010).
[6] F. J. Fattoyev, C. J. Horowitz, J. Piekarewicz and G. Shen, Phys. Rev. C **82**, 055803 (2010); (arxiv:1008.3030).
[7] P. A. M. Guichon, Phys. Lett. **B200** 235 (1988).
[8] K. Saito and A.W. Thomas, Phys. Lett. B **327**, 9 (1994); P.A.M. Guichon, K. Saito, E. Rodionov, and A.W. Thomas, Nucl. Phys. **A601** 349 (1996)
[9] P.K. Panda, C. Providência and D.P. Menezes, Phys. Rev. **C83**, 045801 (2010).
[10] J. Schaffner-Bielich, A. Gal, Phys. Rev. C 62, 034311 (2000); J. Schaffner-Bielich, M. Hanauske, H. Stocker and W. Greiner, Phys. Rev. Lett. 89, 171101 (2002); E. Friedman, A. Gal, Phys. Rep. 452, 89 (2007).
[11] M. Chiapparini, M. E. Bracco, A. Delfino, M. Malheiro, D.P. Menezes, and C. Providncia, Nucl. Phys. A 826, 178 (2009).
[12] Rafael Cavagnoli, Constança Providência, Débora P. Menezes, arXiv:1108.1733v1 [hep-ph].
[13] Prafulla K. Panda, Alexandre M. S. Santos, Débora P. Menezes, Constança Providência, arXiv:1110.4708v1 [nucl-th].
[14] Avraham Gal, Prog. Theor. Phys. Suppl. 186, 270 (2010), and references therein. P. Khaustov *et al.* (BNL E885), Phys. Rev. C 61, 054603 (2000).
[15] P. Danielewicz, R. Lacey and W. G. Lynch, Science **298**, 1592 (2002).
[16] L. W. Chen, C. M. Ko, and B. A. Li, Phys. Rev. Lett. 94, 032701 (2005).
[17] J. M. Lattimer, C. J. Pethick, M. Prakash, and P. Haensel, Phys. Rev. Lett. 66, 2701 (1991).
[18] T. Klähn, *et al.*, Phys. Rev. C 74, 035802 (2006).
[19] A. W. Steiner, J. M. Lattimer, E. F. Brown, Astrophys. J. 722, 33 (2010)
[20] J. R. Oppenheimer and G. M. Volkoff, Phys. Rev. 55, 374 (1939); R. C. Tolman, Phys. Rev. 55, 364 (1939).
[21] Paul Demorest, Tim Pennucci, Scott Ransom, Mallory Roberts, Jason Hessels, Nature 467, 1081 (2010).
[22] M. Prakash, I. Bombaci, M. Prakash, P. J. Ellis, J.M. Lattimer and R. Knorren, Phys. Rep. **280**, 1 (1997).

Journal of Physics: Conference Series **342** (2012) 012009 doi:10.1088/1742-6596/342/1/012009

The pygmy dipole strength, the neutron radius of ^{208}Pb and the symmetry energy

X. Roca-Maza[1,2], M. Brenna[1,3], M. Centelles[2], G. Colò[1,3], K. Mizuyama[1], G. Pozzi[3], X. Viñas[2] and M. Warda[2,4]

[1]INFN, sezione di Milano, via Celoria 16, I-20133 Milano, Italy
[2]Departament d'Estructura i Constituents de la Matèria and Institut de Ciències del Cosmos, Facultat de Física, Universitat de Barcelona, Diagonal 647, 08028 Barcelona, Spain
[3]Dipartimento di Fisica, Università degli Studi di Milano, via Celoria 16, I-20133 Milano, Italy
[4]Katedra Fizyki Teoretycznej, Uniwersytet Marii Curie–Skłodowskiej, ul. Radziszewskiego 10, 20-031 Lublin, Poland

E-mail: `xavier.roca.maza@mi.infn.it`

Abstract. The accurate characterization of the nuclear symmetry energy and its density dependence is one of the outstanding open problems in nuclear physics. A promising nuclear observable in order to constrain the density dependence of the symmetry energy at saturation is the neutron skin thickness of medium and heavy nuclei. Recently, a low-energy peak in the isovector dipole response of neutron-rich nuclei has been discovered that may be correlated with the neutron skin thickness. The existence of this correlation is currently under debate due to our limited experimental knowledge on the microscopic structure of such a peak. We present a detailed analysis of Skyrme Hartree-Fock (HF) plus random phase approximation (RPA) predictions for the dipole response in several neutron-rich nuclei and try to elucidate whether models of common use in nuclear physics confirm or dismiss its possible connection with the neutron skin thickness. Finally, we briefly present theoretical results for parity violating electron scattering on ^{208}Pb at the conditions of the PREx experiment and discuss the implications for the neutron skin thickness of ^{208}Pb and the slope of the symmetry energy.

1. Introduction

Experimental studies [1, 2, 3] on the low-energy isovector dipole response or Pygmy Dipole Resonance (PDR) in neutron rich nuclei are of crucial relevance because they determine reaction rates in the r-process [4] and, in addition, the PDR has been related to the neutron skin thickness of the studied nuclei in Refs. [5, 6, 7]. However, the existence of this correlation is currently under debate [8] due to our limited experimental knowledge of the microscopic structure of such a peak. If the underlying dynamics giving rise to the PDR is, eventually, confirmed to be strongly correlated to the formation of a neutron skin in neutron-rich nuclei, the density dependence of the nuclear symmetry energy at saturation will be immediately constrained [9]. This is of broad interest since the symmetry energy and its density dependence impact on a variety of physical systems, such as the composition and structure of the crust in a neutron star [10, 11], the neutron skin thickness of a heavy nucleus [9, 12, 13], atomic-parity violation [14] and heavy ion collisions [15].

For all these reasons, we have performed in Ref. [16] a detailed analysis of Skyrme Hartree-Fock (HF) plus random phase approximation (RPA) predictions for the isovector and isoscalar dipole response in ^{68}Ni, ^{132}Sn and ^{208}Pb nuclei —representative of different mass regions— in order to elucidate the nature and possible connection of the PDR with the slope of the symmetry energy. The strategy adopted to understand if such a connection may exist, is very simple. Giant resonances are collective excitations of atomic nuclei that have allowed, in the past, to determine some nuclear saturation properties such as the nuclear incompressibility (strongly related to the Giant Monopole Resonance [17]), the effective nucleon mass (which affects the Giant Quadrupole Resonance [18]), or the nuclear symmetry energy at sub-saturation density (which act as a restoring force in the Giant Dipole Resonance [19]). Hence, collective phenomena inform us about general properties of the nuclear effective interaction and all realistic models should predict such a collectivity even though some differences on details may appear. The additional fact that the energy weighted sum rule for the PDR has been correlated with the L parameter [6], defined as $L = 3\rho_0 \left.\frac{\partial c_{\rm sym}(\rho)}{\partial \rho}\right|_{\rho_0}$ where $c_{\rm sym}(\rho)$ is the symmetry energy, ρ the baryon density and ρ_0 the nuclear saturation density, has motivated our selection of studied interactions. Specifically, we use three Skyrme interactions widely used for nuclear structure calculations and that differ in their predictions of the L parameter to study the collectivity displayed by the RPA states giving rise to the pygmy dipole strength, or *RPA-pygmy* state, as well as its possible relation with the neutron skin thickness. Based on our experience, we present here the common features found in different mass regions by using ^{208}Pb as a representative example.

Currently, the PREx collaboration [20] aims to determine the neutron radius of ^{208}Pb within a 1% error by parity violating electron scattering (PVES) [21]. Such a measurement is very important for three basic reasons. First, it measures the neutron distribution in a heavy nucleus free from most of the strong interaction uncertainties. Second, it paves the way for further measurements of neutron densities by PVES [23, 24]. Third, it may allow one to derive a significant constraint on the L parameter [22, 25] and, therefore, it may help in constraining the isovector channel of the nuclear effective interaction. Here, we will shortly discuss mean-field model predictions for PVES at the kinematics of PREx [22].

The work is organized as follows. In Sec. 2 we briefly present the basic formalism employed in our analysis. For further information we address the reader to Refs. [21, 26]. In Sec. 3, we show the main results of our works [16, 22]. Finally, our conclusions are laid in Sec. 4.

2. Formalism

2.1. Random Phase Approximation

The RPA method is well-known from textbooks [26]. In short, once the HF equations are solved self-consistently for the given Hamiltonian, we build accordingly the residual interaction —considered to be a small perturbation of the HF mean-field potential— and, then, we solve the RPA coupled equations by means of the matrix formulation. The important quantities to define for our study are the following. The main one is the reduced transition strength or probability,

$$B(EJ, \tilde{0} \to \nu) \equiv \left| \sum_{ph} A_{ph}(EJ, \tilde{0} \to \nu) \right|^2 = \left| \sum_{ph} \left(X_{ph}^{(\nu)} + Y_{ph}^{(\nu)} \right) \langle p || \hat{F}_J || h \rangle \right|^2 \quad (1)$$

where $A_{ph}(EJ, \tilde{0} \to \nu)$ is the reduced amplitude, $|\tilde{0}\rangle$ is the RPA ground state, $|\nu\rangle$ is a generic RPA excited state, J is the angular momentum carried by the operator, $\hat{F}_{JM}$, meant to modelize the experimental probe and $\langle p||\hat{F}_J||h\rangle$ is the reduced matrix element of such an operator between a hole (h) state (occupied state) and a particle (p) state (unoccupied state). The sum of all ph states that contribute to an RPA transition is weighted by the X^ν and Y^ν amplitudes, eigenvectors of the RPA secular matrix [26]. For further details we also refer to [27].

2.2. Parity Violating Elastic Electron Scattering

Parity violating electron-nucleus scattering (PVES) probes neutrons in a nucleus via the electroweak interaction [28, 29]. Electrons interact with the protons and neutrons of the nucleus by exchanging a photon or a Z^0 boson. The former mainly couples to protons while the latter basically couples to neutrons. In the experiment, one measures the parity-violating asymmetry,

$$A_{pv} \equiv \frac{\frac{d\sigma_+}{d\Omega} - \frac{d\sigma_-}{d\Omega}}{\frac{d\sigma_+}{d\Omega} + \frac{d\sigma_-}{d\Omega}}, \tag{2}$$

where $d\sigma_\pm/d\Omega$ is the elastic electron-nucleus differential cross section for incident electrons with positive and negative helicity states. For a realistic calculation of the parity violating asymmetry, we solve the Dirac equation via the distorted wave Born approximation (DWBA) [30] where the main input are the electric and weak charge distributions of the studied target [22, 25].

3. Results

3.1. Low-energy dipole response of ^{208}Pb

In our recent work [16] we have studied in detail the dipole response of ^{68}Ni, ^{132}Sn and ^{208}Pb by means of the formalism explained in Section 2.1 and by using different Skyrme interactions. Here we will present the case of ^{208}Pb as a representative example of the common trends found in different mass regions.

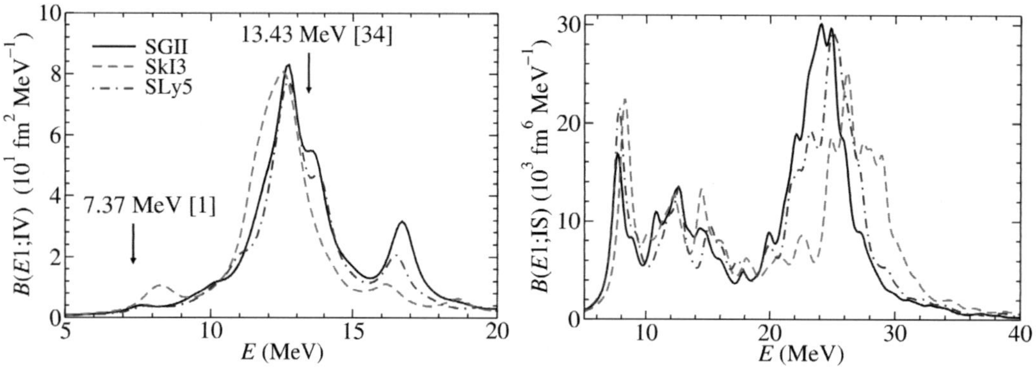

Figure 1. Strength function calculated by convoluting the corresponding reduced transition probability [Eq. (1)] with a Lorenzian of 1 MeV width for the isovector (left) and isoscalar (right) dipole response of ^{208}Pb as a function of the excitation energy. In both figures the predictions of SGII, SkI3 and SLy5 are depicted. Black arrows indicate the experimental centroid energies for the PDR (E = 7.37 MeV) [1] and for the IVGDR (E = 13.43 MeV) [34].

In Fig. 1 we show the isovector (IV: left panel) and isoscalar (IS: right panel) averaged strength functions for the dipole in ^{208}Pb predicted by the Skyrme models SGII [31] with an $L = 37.6$ MeV, SLy5 [32] with an $L = 48.3$ MeV and SkI3 [33] with an $L = 100.5$ MeV as a function of the excitation energy. Experimental data for the centroid energies of the PDR [1] and the IV Giant Dipole Resonance (GDR) [34] are also depicted (black arrows). The theoretical predictions lie within the experimental error and, therefore, the Skyrme-HF plus RPA approach may constitute a good starting point for a detailed analysis of the microscopic structure of the PDR in those nuclei. An interesting feature that can be also observed in different mass regions [16] is the ratio between the low- and high-energy strength exhausted by the IS peaks: of the same order; and

Journal of Physics: Conference Series **342** (2012) 012009 doi:10.1088/1742-6596/342/1/012009

by the IV peaks: one order of magnitude. This already indicates the importance of the probe used to excite a certain RPA state and reveals also its nature. In other words, the IS dipole operator excite more efficiently the *RPA-pygmy* state than the IV dipole operator. This means that a perfect probe for exciting such a state would be mostly isoscalar. Moreover, we see from both figures that as the predicted value for the L parameter increases —in going from SGII to SLy5 and, finally, to SkI3, the low-energy peak is shifted to larger excitations energies and to larger strengths, in perfect agreement with Ref. [6].

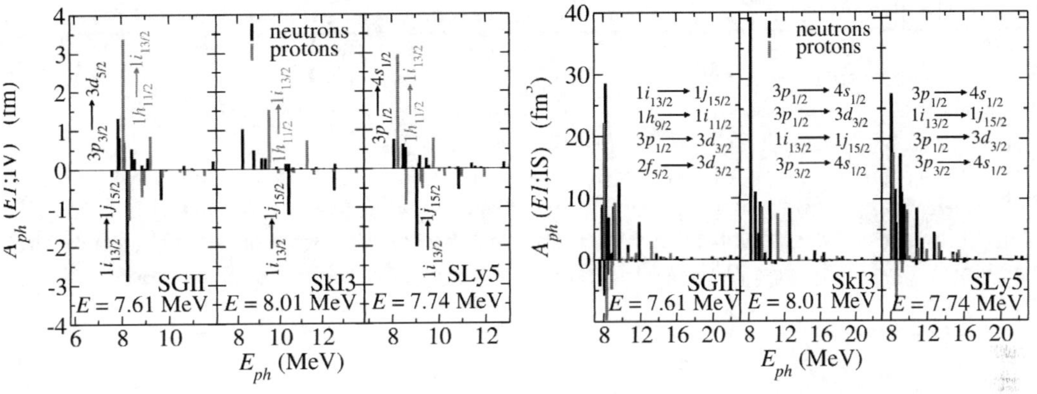

Figure 2. Neutron (black) and proton (red) *ph* contributions to the isovector (left) and isoscalar (right) reduced amplitude [Eq. (1)] corresponding to the *RPA-pygmy* state of ^{208}Pb (RPA excitation energy is indicated) predicted by the different models as a function of the *ph* excitation energy. The single particle levels involved in the most important *ph* transitions are indicated.

In Fig. 2, we show the *ph* contributions to the IV (left) and IS (right) reduced amplitudes [see Eq. (1)] corresponding to the different *RPA-pygmy* states of ^{208}Pb predicted by the different models as a function of the *ph* excitation energy. Two results —common also to other mass regions— arise from the analysis of this quantity. First, while the isoscalar reduced amplitude ($A_{ph}(E1, IS)$) is formed by several neutron *ph* transitions adding coherently (same sign) and only few states are contributing destructively, the isovector reduced amplitude ($A_{ph}(E1, IV)$) is formed by few neutron and proton *ph* transitions contributing with different signs, i.e., non-coherently. Second, the most relevant *ph* transitions in the isoscalar dipole response of ^{208}Pb correspond to excitations of the outermost neutrons that, as a consequence, dominate the dynamics in the low-energy region (see [16]). For the case of the isovector dipole response the situation is model dependent. However, the most relevant neutron *ph* contributions are larger in number than the proton ones and basically due to the outermost neutrons. Therefore, albeit some differences will be present, one may expect a relation between the neutron excess and the IS and IV dipole responses in neutron rich nuclei.

From Fig. 1, we have seen that the low-energy IS and IV dipole responses of ^{208}Pb display, to different degree, a sizeable low-energy peak in the corresponding strength functions (see section III.B of Ref. [16]). This is actually one of the main characteristics one asks to a collective state. The second one is that collective phenomena (or resonances) should display coherence between the contributions of several *ph* transitions to the reduced amplitude. Therefore, by looking at Fig. 2, we can state that while all models support a clear collective character of the low-energy peak in the IS dipole response in ^{208}Pb (also true in other mass regions), the collectivity of its IV counterpart is model dependent.

3.2. Parity violating electron scattering on ^{208}*Pb*

The Lead Radius Experiment (PREx) at the Jefferson Laboratory has recently reported first results for PVES on ^{208}Pb [20]. In this first run statistics were not sufficient in order to achieve the desired accuracy [20]. A second run of PREx has been approved and it is intended to be performed in the future [20, 23].

As stated in the introduction, the relevance of this measurement has motivated our study of the parity violating asymmetry at the PREx kinematics [21, 22]. We have applied the formalism mentioned in Section 2.2, which is presented in more detail in Refs. [21, 22, 29, 30]. In Fig. 3 we display the linear correlation between the parity violating asymmetry and the neutron skin thickness of ^{208}Pb ($\Delta r_{np} = \langle r_n^2 \rangle^{1/2} - \langle r_p^2 \rangle^{1/2}$) as predicted by more than forty mean-field models of very different nature: from non-relativistic such as Skyrme or Gogny models to relativistic such as non-linear Walecka or density dependent meson-exchange and point-couling models. All of them accurately reproduce the charge radius of ^{208}Pb (for further details see Ref. [22]). This high linear correlation allows one to accurately extract the value of the neutron skin thickness of ^{208}Pb without assuming a particular shape for the nucleon spatial distributions. In the right panel of Fig. 3 we show the well-known correlation between the neutron skin thickness in ^{208}Pb and the slope of the symmetry energy predicted by the considered mean-field models. Therefore, PVES can supply new constraints on the value of the L parameter.

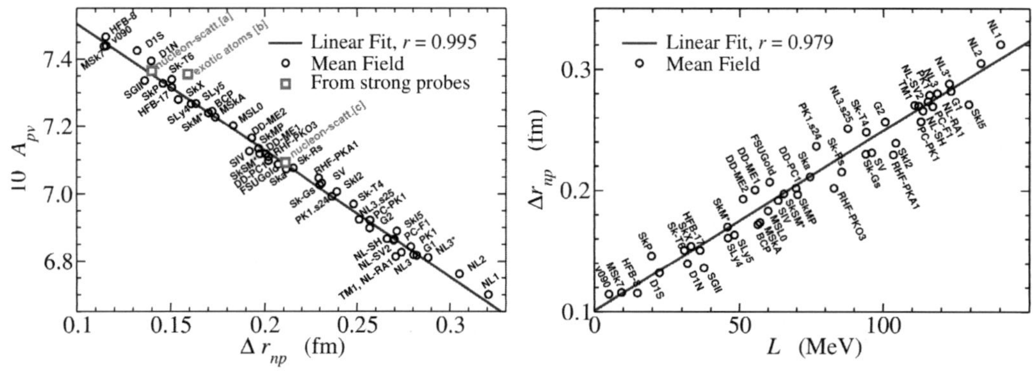

Figure 3. Left panel: parity violating asymmetry (DWBA) in ^{208}Pb for 1.06 GeV electrons at 5° scattering angle as a function of its neutron skin thickness. MF results (black circles) —references to all these interactions can be found in [22]— and calculations form the neutron densities deduced from experiment [a] [35], [b][36] and [c] [37] (red squares). Right panel: Neutron skin thickness of ^{208}Pb as a function of L as predicted by the same MF interactions shown in the left panel.

4. Conclusions

We have exemplified some of the common features found in the dipole response of different nuclei [16] by showing our results for the case of ^{208}Pb. In particular, the low-energy peak in both the isoscalar and isovector dipole responses is shifted to larger excitation energies and display larger values of the strength function as the value of the L parameter increases. We have also seen that the *RPA-pygmy* state can be more efficiently excited by an isoscalar probe than by an isovector probe. This indicates the dominant isoscalar character of such a state. We have demonstrated that the low-energy peak in the isoscalar dipole response is a collective mode —it is formed by several *ph* contributions adding coherently and giving a contribution to the reduced transition strength comparable to that of the isoscalar giant dipole resonance. This is found to be opposite to what happens to its isovector counterpart where its collectivity depends on the interaction. Finally, the dynamics in the low-energy region in the isoscalar dipole response of neutron-rich nuclei is clearly dominated by the outermost neutrons —those that form a neutron skin.

Our analysis of PVES applied to the conditions of the PREx experiment predicts a high-quality correlation between the parity violating asymmetry and the neutron skin thickness of ^{208}Pb [22]. The results suggest that one will be able to extract significant constraints on the slope of the nuclear symmetry energy at saturation if the statistics of PREx are improved.

In conclusion, the recent experimental and theoretical studies of the ground-state and excitation properties of neutron-rich nuclei aim to complement each other and are paving the way for a better knowledge of the isovector channel of the nuclear effective interaction.

Acknowledgments

Work partially supported by grants CSD2007-00042, FIS2008-01661 and 2009SGR-1289 (Spain) and DEC-2011/01/B/ST2/03667 from NCN (Poland).

References

[1] Ryezayeva N *et al.* 2002 *Phys. Rev. Lett.* **89**, 272502
[2] Adrich P *et al.* 2005 *Phys. Rev. Lett.* **95** 132501
[3] Wieland O *et al.* 2009 *Phys. Rev. Lett.* **102** 092502
[4] Paar N, Vretenar D, Khan E and Colò G 2007 *Rep. Prog. Phys.* **70** 691
[5] Klimkiewicz A *et al.* 2007 *Phys. Rev. C* **76** 051603(R)
[6] Carbone A *et al.* 2010 *Phys. Rev. C* **81** 041301
[7] Inakura Tsunenori, Nakatsukasa Takashi and Yabana Kazuhiro 2011 *Phys. Rev. C* **84** 021302(R)
[8] Reinhard P G and Nazarewicz W 2010 *Phys. Rev. C* **81** 051303
[9] Brown B A 2000 *Phys. Rev. Lett.* **85** 5296
[10] Horowitz C J and Piekarewicz J 2001 *Phys. Rev. Lett.* **86** 5647
[11] Roca-Maza X and Piekarewicz J 2008 *Phys. Rev. C* **78** 025807
[12] Centelles M, Roca-Maza X, Viñas X and Warda M 2009 *Phys. Rev. Lett.* **102** 122502
[13] Warda M, Viñas X, Roca-Maza X and Centelles M 2009 *Phys. Rev. C* **80** 024316
[14] Sil T, Centelles M, Viñas X and Piekarewicz P 2005 *Phys. Rev. C* **71** 045502
[15] Tsang M B *et al.* 2009 *Phys. Rev. Lett.* **102** 122701
[16] Roca-Maza X, Pozzi G, Brenna M, Mizuyama K and Colò G 2011 Submitted to *Phys. Rev. C.*
[17] Colò G, Van Giai N, Meyer J, Bennaceur K and Bonche P 2004 *Phys. Rev. C* **70** 024307
[18] Reinhard P G 1999 *Nucl. Phys. A* **649**, 305c (1999).
[19] Trippa L, Colò G and Vigezzi E 2008 *Phys. Rev. C* **77**, 061304(R)
[20] Kumar K, Souder P A, Michaels R and Urciuoli G M, spokespersons, http://hallaweb.jlab.org/parity/prex.
[21] Horowitz C J, Pollock S J, Souder P A and Michaels R 2001 *Phys. Rev. C* **63** 025501
[22] Roca-Maza X, Centelles M, Viñas X and Warda M 2011 *Phys. Rev. Lett.* **106** 252501
[23] Proposal to Jefferson Lab PAC 38 "PREX-II: precision parity-violating measurement of the neutron skin of Lead" http://hallaweb.jlab.org/parity/
[24] Ban Shufang, Horowitz Charles J and Michaels R 2011 *J. Phys. G* in press [arXiv:1010.3246v2]
[25] Centelles M, Roca-Maza X, Viñas X and Warda M 2010 *Phys. Rev. C* **82** 054314
[26] Ring P and Schuck P *The Nuclear Many-Body Problem* (Springer-Verlag, New York, 1980).
[27] Colò G *et al.* Submitted to *Comp. Phys. Comm.*
[28] Donnelly T W, Dubach J and Sick I 1989 *Nucl. Phys. A* **503** 589
[29] Horowitz C J 1998 *Phys. Rev. C* **57** 3430
[30] Roca-Maza X, Centelles M, Salvat F and Viñas X 2008 *Phys. Rev. C* **78** 044332
[31] Van Giai N and Sagawa H 1981 *Phys. Lett. B* **106** 379
[32] Chabanat E *et al.* 1998 *Nucl. Phys. A* **643**, 441
[33] Reinhard P G and Flocard H 1995 *Nucl. Phys. A* **584** 467
[34] Berman B L and Fultz S C 1975 *Rev. Mod. Phys.* **47** 713
[35] Hoffmann G W *et al.* 1980 *Phys. Rev. C* **21** 1488
[36] Kłos B *et al.* 2007 *Phys. Rev. C* **76** 014311
[37] Zenihiro J *et al.* 2010 *Phys. Rev. C* **82** 044611

Journal of Physics: Conference Series **342** (2012) 012010 doi:10.1088/1742-6596/342/1/012010

Intense infrared lasers and laboratory astrophysics

Luis Roso

Centro de Laseres Pulsados, CLPU, Patio de Escuelas 1, 37008 Salamanca, Spain

E-mail: roso@usal.es

Abstract. Electrons accelerate due to the huge electric field of the laser itself, and so they can generate very particular plasmas because the ionization process occurs in a few femtosecond, or even in a fraction of a femtosecond. Lasers can be focused now to intensities beyond $10^{22} W/cm^2$ and there are projects to arrive up $10^{26} W/cm^2$. Electric fields of the laser arrive now to $10^{14} V/cm$ and magnetic fields reach the Megatesla. This is a monster density of electromagnetic energy, so that we are close to obtain light denser than matter. In this respect it is very convenient to observe that the well known Einstein's energy - mass equation, $E = mc^2$, can be rewritten for laser light $I = \rho c^3$, I being the laser intensity and ρ the equivalent density. There are several PW lasers around the world, in operation or in construction, and one of them is going to be at Salamanca's CLPU.

1. Introduction

Extreme plasma physics needs extreme tools and present day lasers can be considered as such extreme tool. The necessity of extreme lasers was clear soon after the discovery of the lasers -if not before-. By mid-eighties, a group of researches at the Laser Laboratory at the University of Rochester (Rochester, NY), lead by Gerard Mourou, discovered a very clever trick to boost the laser peak power. This was the launching of the ultraintense laser technology. This new technique was soon known as Chirped Pulse Amplification, since it relies in the introduction of chirping in the broadband laser pulses before amplification [1]. Chirped Pulse Amplification is now a well developed technology [2]. It has allowed the most intense lasers ever dreamed of [3].

Now with those lasers monster pressures, huge accelerations, and electric and magnetic field without equivalent on Earth. Moreover we try to justify that those lasers differ very much of the standard concept of a laser. First, they have nothing to do with monochromatic light. Those lasers are short time (femtosecond) and thus they are broadband. Second, they are not really light beams, but light bullets. A 30 femtosecond pulse, for example corresponds -for 800 nanometer wavelength- to 10 microns. So when focused, this laser delivers something closer to a light bullet than to a light beam [4].

2. Present status of light-mater interaction phenomenology.

It is important to concentrate the peak power over a very small focal spot, because most of the phenomenology depends on the electric field at focus [5]. The phenomenology of the interaction of such lasers with atoms, molecules and nuclei is very rich and surprising. Our understanding of the light-mater phenomenology has been very rapidly changing over the last decades. We review now the most important milestones.

Below 10^8 W/cm^2, approximately, the phenomenology is the standard Einstein photoelectric effect, in the sense that the photon is so short in energy to ionize. However, from 10^{10} W/cm^2, there are so many photons involved in the ionization that several of them can bring the necessary energy for releasing the electron, leading to multiphoton ionization. One of the main rules of Einsteins photoelectric effect, the need of a photon more energetic than the binding energy of the electron now fails and ionization can occur with the absorption of a number of photons, not necessarily the minimum number of photons to arrive to the continuum, but more of them (in that case, the effect is known as above threshold ionization). This is the prevailing phenomenology until the electric field itself begins to play a role and suppress the Coulomb barrier that bounds the electron to the atomic nucleus. Beyond 10^{16} W/cm^2, electrons are simply released over the ionization barrier by something similar to the DC Stark effect. From this point on, ionization is simply due to the electric field strength, and dynamics is completely different from the multiphoton regime.

To have an idea of the strength of the field it is worth to remember the definition of the atomic unit of intensity. Atomic units are determined by $m = e = \hbar = 1$, $\hbar$, being the Planck constant, m the electron mass, and e the absolute value of the electron charge. In atomic units the speed of light is $c = 137$. The atomic unit of intensity is 3.57 10^{16} W/cm^2, an intensity that has been achieved for more than two decades. This intensity of the laser field corresponds to an electric field amplitude of 5.14 $10^9 V/cm$. Just by the definition of atomic unit, the atomic unit of electric field this is the field experienced at a distance equal to one atomic unit (one Bohr radius) of a proton, i.e. is the electric field that an electron in the ground state of the hydrogen atom feels. This means that beyond the atomic unit of intensity electrons-laser coupling is stronger that electron-nucleus coupling. So beyond atomic unit of intensity electrons are suddenly released from atoms by over the barrier ionization and move freely driven by the laser field. Thus generating instantaneously -in a few femtoseconds- a strongly driven plasma.

3. Laser acceleration.

As intensity increases, atoms have no meaning, and we must consider just a plasma of electrons and positive ions. Not only valence electrons are ionized, but inner shells can arrive to the barrier suppression regime. Two orders of magnitude more and the magnetic of the laser field begins to be relevant for the dynamics of the driven electrons. For the Ti:Sapphire wavelength (800 nanometer) ionized electrons travel at relativistic speeds [5] at intensities beyond 10^{18} W/cm^2. A relevant feature of the strong laser field driven electron is that it starts at rest, accelerates to a large speed in a one or a few femtoseconds. The acceleration is extremely high and so radiation effects may become important.

There are also some laboratories working in laser acceleration with far infrared lasers, as the 10 micrometers CO_2 laser [6]. But such long wavelengths imply other unavoidable fundamental problems: the focal spot minimum size is governed by diffraction and so is proportional to the wavelength. On the other side of the spectra, ultraviolet lasers have fields oscillating too fast for a convenient acceleration in one cycle. As a result, near infrared seems nowadays to be the best option for acceleration.

As we jump in the intensity ladder to values of the order of 10^{21} W/cm^2, protons are also relativistically driven. As protons have a larger mass they need a higher intensity to become relativistic. As intensity increases other ions are also relativistically driven. In all cases, obviously the important parameter is the charge/mass relation.

4. Relativistic plasmas.

In the same way a laser accelerates a free particle, it accelerates a collection of charged particles a plasma. When a laser is focused at those relativistic intensities on a target, it generates a plasma of relativistic electrons and if the intensity is large enough- also of relativistic ions.

Journal of Physics: Conference Series **342** (2012) 012010 doi:10.1088/1742-6596/342/1/012010

The plasma appears in a few femtoseconds and thermatization is at those time scales a slow process. Released electrons moving close to the light speed have no time to feel their electrostatic repulsion and keep plasma densities close to solid densities for a few femtoseconds thus generating huge electrostatic fields that in turn can help in the accelerations process. As a consequence, new plasma effects such as wakefield acceleration for tens of femtoseconds pulses- and bubble accelerations for few femtoseconds pulses- are now under study. We must also indicate that it is extremely difficult to measure the intensity at such frontier region. We can assume that a petawatt laser with a good wavefront quality can arrive to 10^{23} W/cm^2, without great difficulty. Observe that a petawatt (10^{15} watt) focused into a spot of one squared micrometer corresponds to that intensity. However many times the best wavefront quality is achieved with a somewhat not so extreme intensity, and the reported records, close to 10^{23} W/cm^2, correspond to lasers of multi-hundred terawatt power [7] . This 10^{23} W/cm^2 number can be considered as today's world record of intensity. It is feasible with titanium:sapphire lasers (800 nanometer) and about 30 femtoseconds pulse duration. This intensity corresponds to 8.6 10^{12} V/cm, for the electric field amplitude, and a magnetic field amplitude of 2.8 10^6 Tesla, which are extraordinary numbers. Of course, those fields are oscillating fields with a 2.6 femtosecond period, so those maximum amplitudes last one femtosecond (imagine if those values were with static fields). But at those extreme fields even a fraction of a femtosecond is enough to induce extreme effects.

5. Density equivalent

To convince the reader that a laser pulse well focused correspond to a monster concentration of energy we propose the following analogy. A 30 femtosecond laser pulse is a veil of light just 10 microns thick moving at the speed of light. If this is focused to a spot size of, for example, 1 squared micron, we have all the pulse energy concentrated on a volume of 10 cubic microns. A very small volume. But this is the key feature of lasers, we are now able to concentrate energy in a never before dreamed way. More precisely if ρ is the mass density (mass divided by volume) then the energy density is $\rho\ c^2$ for mass. In the case of a laser focused to a given intensity I, it is straightforward to show that the energy density is given by I/c. So we can consider that each intensity corresponds to a mass density, understanding that this is just an analogy to understand the meaning of such extreme fields. The density to intensity relation is therefore $I = \rho\ c^3$. This relation is very basic and illuminating. Always has been considered that matter is a huge concentration of energy, and when just a tiny fraction of that energy can be released (as in nuclear reactions) it has important consequences. Light is energy by itself, since it is electromagnetic energy. Einstein mass energy relation corresponds approximately to 9 10^{13} Joule/gram. This is a very important quantity of energy because we consider one cubic centimeter of water, for example. If instead of this volume, we would consider just one cubic micrometer of water, then this energy is just 90 J, an important energy but maybe soon achievable with a laser.

What we learn from Table 1 is that present day lasers are close to water density equivalent. According to the evolution of the technology it is expected to arrive in a couple of years to intensities of the order of 10^{25} to 10^{26} W/cm^2. Although this is just a didactic analogy, it shows that state of the art CPA lasers are a fabulous tool that provides a huge concentration of electromagnetic energy, relevant for many applications and without precedent on Earth.

6. Is there a limit in the laser intensity?

A laser beam is a collection of photons, since photons are bosons then it can be possible to place infinite photons in the same state and thus there are no limits for the intensity of any laser. This seemed just an academic question for many decades, but maybe we are close to

Journal of Physics: Conference Series **342** (2012) 012010 doi:10.1088/1742-6596/342/1/012010

Table 1. Some typical values of the density of mass and their translation into energy density according to the Einstein energy-mass conversion equation. At the right column we can see their intensity equivalent .

	density gr/cm^3	energy density $J/\mu m^3$	intensity W/cm^2
air	0.001	0.09	2,69 10^{21}
today's-lasers	0.037	3.40	10^{23}
water	1	90	2,69 10^{24}
sun-core	100	9000	2,69 10^{26}
Schwinger-limit	85733	3 10^6	2,3 10^{29}
withe-dwarf	1000000	9 10^7	2,69 10^{30}

experimentally explore this limit. Quantum electrodynamics, QED, is the theory that describes the interaction of electromagnetic fields with charged particles, as well as the creation of pairs of particles in the presence of such fields, and related processes. From the early times of QED it was clear the existence of a critical field that can not exceed because of the spontaneous creation of electro-positron pairs. This critical field is now known as the Schwinger limit, since it was clearly identified and studied by this author [8] although it appeared in other previous works. The Schwinger critical field is 1.3 10^{16} V/cm , that corresponds to a laser intensity equal to 2.3 10^{29} W/cm^2 and to a magnetic field of 4.4 10^9 Tesla.

At early QED times the Schwinger limit was just academic, since it was too far from experimental possible sources. Now this limit is still far but not too far and a lot of recent research activity is trying to understand much more precisely the Schwinger limit [9] , and its onset. Close to the critical field, electron-positron pairs can appear from vacuum and vacuum thus becomes unstable since the number of pairs is so large that they take a significant fraction of the laser pulse energy.

Since we are entering the region where light can generate pairs from vacuum, a lot of new questions are appearing those days. Moreover, two co-propagating high energy-photons cannot generate a particle pair since momentum-energy conservation has to be violated. In the case of the laser this is true, but now one million photons combine to give a pair, and momentum-energy conservation can be addressed in different practical ways: - If the laser beam is divided into two parts that collide head on. In that case pair creation process is maximized. This is closer to standard photon-photon collision with the peculiarity that the centre of mass energy of the photon-photon collision is a few eV, so millions of photons must act collectively to generate a pair. A number of studies are now under progress to design experiments of photon-photon scattering in vacuum [10] .

- A focused beam implies a convergent wavefront and thus all photon momenta are not strictly parallel. So pairs can also appear in a relatively efficient way.

- A last strategy to check the QED limit is to use just a collimated beam and to introduce a counter-propagating accelerated electron. The Lorentz boost from the lab frame to the approaching electron frame enhances the efficiency of pair production [11].

There are thus many strategies under progress to analyse the decay of the quantum vacuum. Also there are many uncertainties, that now are very relevant because they can imply design of very expensive near future experiments. Two of the most relevant are:

- Are there charged particles with a mass smaller than the electron mass? If so, since the Schwinger critical field scales with mass, pairs of these hypothetical particles would appear before electron-positron pairs.

- Pair creation process is a tunnelling process and tunnelling dynamics still is not very understood in many aspects. In this case, the generated particles appear inside a huge electromagnetic field, and therefore they suffer an intense acceleration just at the moment they are created. Most of the theories indicate that the pair is first generated and once it is real both individual particles are accelerated. However it could also be possible that the energy for the pair creation is not just the rest energy but the energy they need to gain to be inside the field (the so called ponderomotive energy). In that case the Schwinger limit will be shifted to higher intensities.

Therefore, close to the Schwinger limit there should be a fundamental limit of the intensity attainable with a high power laser. It is not clear if this limit will occur before or after the Schwinger critical field, but it is clear that something will happen and that high intensity studies are necessary. There are some efforts as ELI [12] to build in the future hundred Petawatt lasers that will allow experimental work on this. At 10^{26} W/cm^2, if not before, most theories indicate that vacuum is non-linear and that virtual pair production will have some effects. Maybe such intense lasers will propagate in vacuum at a speed lower that c?

Having a conceptual upper limit for laser intensity, even when the laser is just loosely focused, is an interesting basic physical problem that needs a lot of future theoretical research. But such necessity of new modeling does not end with the existence of an upper limit. It is clear that pairs will appear and that those pairs will be violently accelerated. There are some clues that indicate that the acceleration of electrons in such extreme lasers can be relevant for general relativity and it is expected that such electrons self-generated, or injected- close to the Schwinger limit will radiate in a similar way as in the proximity of a black hole. This kind of radiation, the so-called Unruh radiation is now subject of intensive study, and fundamental experiments to measure it are under design [13].

7. Applications in astrophysics

Probably the existence of such fundamental questions is enough to trigger new research in ultraintense lasers and to justify the important economical efforts necessary to build them in the next decade. Moreover there are a lot of applications of such future lasers. Among them, the most relevant application is particle acceleration. Lasers are starting to be considered as the new paradigm for particle acceleration because a laser can accelerate a particle to GeV or maybe TeV energies in a few millimeters distance.

Laser accelerated plasmas are a new experimental field that can be relevant for several applications including the modeling of astrophysical phenomena. There are a number of publications devoted to the laboratory model of astrophysical processes [14]. Among them, there is a number of nonlinear plasma effects that can be clarified with lasers. There are also some problems on relativistic astrophysics, as cosmic ray acceleration, and as the interaction of electromagnetic waves with relativistic plasmas that can be clarified with the use of ultraintense lasers. Also relativistic solitons and vortices and colisionless shock waves can be studied with lasers.

Being a bit more optimistic, future lasers can generate so violent accelerations that can propose new specific tests to general relativity. The experimental search of Unruh radiation in laser accelerated plasmas can help to the validation of some models of black holes.

Anyhow, extreme laser modeling of astrophysical plasmas will take some time, probably decades to be mature. In between lasers are becoming relevant to study nucleogenesis and fusion reactions similar to the Sun's interior, and the experimentation with dense plasmas is advancing very quickly.

8. Conclusions.

Ultraintense lasers are evolving as a conceptually new tool to explore the quatum vacuum. What can be learned from such experiments is not clear, since the experiments themselves are

Journal of Physics: Conference Series **342** (2012) 012010 doi:10.1088/1742-6596/342/1/012010

not clearly designed yet. Anyhow experiments that seem just it gedanken experiments a few years ago, can be considered now as feasible for the next decade. Existing Petawatt lasers constitute one of the frontiers of physics and technology with important applications. But a future generation of multiPetawatt lasers will be able to analyze very fundamental physical problems. Most theoretical models suggest that there is an upper limit of the intensity attainable for a laser, the Schwinger limit. What is more relevant is that some designed lasers are not too far from that fundamental upper limit. The most important question now is to understand what will be the first features of the Schwinger limit that can be experimentally seen.

Acknowledgments
We acknowledge support from Spanish Ministerio de Ciencia e Innovacion through the Consolider Program SAUUL (CSD2007-00013) and the research project FIS2009-09522, and from the EC Seventh Framework Programme (LaserLab Europe, grant agreement number 228334) and ELI Preliminary Phase. The Pulsed Laser Centre (Centro de Laseres Pulsados, CLPU) is a new research facility that has been created as a Consortium of the Spanish Ministry of Education and Science, the Regional Government of Castilla y Leon and the University of Salamanca, as part of the implementation of the Spanish Scientific Infrastructures Roadmap. CLPU is a facility giving beam access to the domestic and international scientific community.

References

[1] D Strickland and G Mourou, Compression of amplified chirped optical pulses, Optics Comm. 55, 447 (1985); D Strickland and G Mourou, Compression of amplified chirped optical pulses, Optics Comm. 56, 219 (1986); P Maine, D Strickland, P Bado, M Pessot, and G Mourou, Generation of ultrahigh peak power pulses by chirped pulse amplification, IEEE Journal on Quantum Electronics 24, 398 (1988); M Pessot, P Maine, and G Mourou, 1000 times expansion/compression of optical pulses for chirped pulse amplification, Opt. Comm. 62, 419 (1987).

[2] S Backus, Ch G. Durfee III, M M. Murnane, and H C. Kapteyn, High Power Ultrafast Lasers, Rev. of Scientific Instr. 69, 1207 (1998) .

[3] Thomas Brabec and Ferenc Krausz, Rev Mod Phys, 72, 545 (2000)

[4] G Mourou, Ch P J Barty, and M D Perry, Ultrahigh-Intensity Lasers: Physics of the Extreme on a Tabletop, Phys. Today 51, January 22 (1998)

[5] G Mourou, T Tajima, and S V Bulanov, Optics in the relativistic regime, Rev. Mod. Phys., 78, 309 (2006).

[6] N. E. Andreev and S. V. Kuznetsov, and I. V. Pogorelsky, Monoenergetic laser wakefield acceleration, Phys. Rev. ST Accel. Beams 3, 021301 (2000)

[7] V. Yanovsky, V. Chvykov, G. Kalinchenko, P. Rousseau, T. Planchon, T. Matsuoka, A. Maksimchuk, J. Nees, G. Cheriaux, G. Mourou, and K. Krushelnick, Ultra-high intensity- 300-TW laser at 0.1 Hz repetition rate, Optics Express, 16, 2109 (2008)

[8] J Schwinger, Gauge Invariance and Mass. II, Physical Review, 128, 2425 (1962)

[9] A. M. Fedotov and N. B. Narozhny, G. Mourou, G. Korn, Limitations on the Attainable Intensity of High Power Lasers, Phys Rev Lett, 105, 080402 (2010)

[10] Daniele Tommasini, Albert Ferrando, Humberto Michinel, and Marcos Seco, Detecting photon-photon scattering in vacuum at exawatt laser, Phys Rev A 77, 042101 (2008)

[11] The main features of the generation of a pair of particles by laser fields have been already demonstrated in the SLAC-144 experiment, back in 1999. See, C Bamber, S J Boege, T Koffas, et al., Studies of nonlinear QED in collisions of 46.6 GeV electrons with intense laser pulses, Phys. Rev. 60, 092004 (1999).

[12] The official information on the Extreme Light Infraestructure can be found at http://www.extreme-light-infrastructure.eu/

[13] P Chen, and T Tajima, Testing Unruh Radiation with Ultraintense Lasers, Phys. Rev. Lett. **83**, 256 (1999)

[14] See, for example, B. Remington, R. Drake, and D. Ryutov, Rev. Mod. Phys. 78, 755 (2006); S. V. Bulanov, T. Zh. Esirkepov, D. Habs, F. Pegoraro, and T. Tajima, Eur. Phys. J. D; and references therein. 55, 483 (2009).

Journal of Physics: Conference Series **342** (2012) 012011 doi:10.1088/1742-6596/342/1/012011

Fully antisymmetrised dynamics for bulk fermion systems

Klaas Vantournhout and Hans Feldmeier

GSI Helmholtzzentrum für Schwerionenforschung GmbH
Planckstraße 1
64291 Darmstadt, Germany

E-mail: k.vantournhout@gsi.de

Abstract. The neutron star's crust and mantel are typical examples of non-uniform bulk systems with spacial localisations. When modelling such systems at low temperatures, as is the case in the crust, one has to work with antisymmetrised many-body states to get the correct fermion behaviour. Fermionic molecular dynamics, which works with an antisymmetrised product of localised wave packets, should be an appropriate choice. Implementing periodic boundary conditions into the fermionic molecular dynamics formalism would allow the study of the neutron star's crust as a bulk quantum system. Unfortunately, the antisymmetrisation is a non-local entanglement which reaches far out of the periodically repeated unit cell. In this proceeding, we give a brief overview how periodic boundary conditions and fermionic molecular dynamics can be combined without truncating the long-range many-body correlation induced by the antisymmetry of the many-body state.

1. Introduction

Neutron stars represent one of the densest forms of matter in the observable universe. As an "outgrown nucleus", they confine approximately 10^{57} baryons inside a radius of roughly 10 km. This leads, in the centre, to densities expected to be as large as five to ten times the normal nuclear saturation density n_0. In the outer layers of a neutron star, the outer and inner crust which represents only a tiny fraction of the total mass, the densities are on the average below n_0. Nonetheless, it is associated with various astrophysical phenomena such as the cooling of neutron stars or neutron star glitches [1]. A detailed knowledge of crustal matter will help to improve our understanding of these phenomena.

The neutron star's crust consists of protons, neutrons and electrons that organise themselves as a result of the combined effect of the short-range nuclear attraction and the long-range Coulomb repulsion. In the outer crust, i.e. the lower-density regions, nucleons bind to form nuclei that are placed in a Coulomb lattice embedded in an electron gas. In the inner crust, the neutron-drip density ($\approx 4 \times 10^{11}\text{g/cm}^3$ [2]) is surpassed giving rise to neutron rich nuclei placed on a similar Coulomb lattice, immersed in a gas of neutrons and electrons. In the more dense layers of the crust ($n_0/3$ to $n_0/2$ [3]), the spherical nuclei outgrow the lattice spacing. In this crust-core interface, i.e. the mantle of the neutron star, it is believed that the system balances on a subtle interplay between the nuclear surface energy and the Coulomb energy of the neutron-proton-electron system. This results in a multitude of competing quasi-ground states

Journal of Physics: Conference Series **342** (2012) 012011 doi:10.1088/1742-6596/342/1/012011

with similar energies, often having quite different matter distributions where the system can cool down to. These complex-shaped foam like structures are referred to as "nuclear pastas" [1, 4, 5].

2. Fermionic molecular dynamics

The crusts and especially the mantel of a neutron star are excellent examples of bulk fermion matter where spatial localisations play an important role. Many computational techniques have been used to investigate these. These range from the liquid drop model over Hartree Fock to Monte Carlo and molecular dynamics [4, 6, 7, 8, 9, 10, 11]. With a few exceptions, all of these methods study a "single" geometry of the nuclear cluster. This geometry is either explicitly implemented or strongly correlated to the shape of the unit cell and the imposed boundary conditions. However, the implementational or computational constraints should not distort the properties of the macroscopic system. The Monte Carlo and molecular dynamics methods are unbiased with regards to the geometry of the nuclear clusters and allow the study of more realistic systems. As there are many configurations with almost the same energy, the nuclear pasta phases are susceptible to small perturbations stemming from external probes or slight temperature changes. The molecular dynamics picture is appropriate to study such systems and is already successfully used for classical descriptions. From the many existing approaches [12, 13, 14, 15, 16], classical molecular dynamics (CMD) provided a way to study the effect of large density fluctuations on neutrino opacities and evaluated the breaking strain of the neutron star's crust [7, 17]. Quantum molecular dynamics (QMD), on the other hand, identified various pasta structures and evaluated one of the first phase diagrams of the different pasta phases [8].

Disregarding the classical nature of CMD and QMD, their achieved successes in the study of nuclear matter at sub-nuclear densities have given us a first understanding of the crust's bulk properties. However, when studying phenomena at length scales smaller then the de Broglie wavelength of the system's constituents, quantum effects stemming from the wave character of the nucleons predominate. In addition, the statistical many-body correlations related to the system's fermionic content will play a fundamental role. While CMD lacks both the wave character as well as the statistical description, QMD works with density packets representing distinguishable particles. The indistinguishablity of the fermions is mimicked by means of a phenomenological Pauli potential which prevents identical fermions to occupy the same phase space [18]. Another fundamental quantum feature missing in QMD is the dynamical variation of individual spin components and the spreading of the wave packets.

The shortcomings of QMD for fermion systems are remedied in fermionic molecular dynamics (FMD) [15, 16]. While QMD uses a phenomenological Pauli potential to mimic fermion statistics, FMD introduces the fermion many-body states as antisymmetrised products, i.e. Slater determinants, of localised single particle states $|q_p(\boldsymbol{z}_p)\rangle$:

$$|Q(\boldsymbol{z}_1,\ldots,\boldsymbol{z}_A)\rangle = \hat{\mathcal{A}}|q_1(\boldsymbol{z}_1)\rangle \otimes \cdots \otimes |q_A(\boldsymbol{z}_A)\rangle. \tag{1}$$

This way, the Pauli exclusion principle and uncertainty principle are implemented from the beginning and cannot be violated. With a proper choice of the single-particle states $|q_p(\boldsymbol{z}_p)\rangle$ and their variation parameters $\boldsymbol{z}_p$, concepts as average phase-space position, spin, isospin and spreading can be introduced. While the equations of motion of the many-body fermion system can be obtained by means of the time-dependent variational principle, structure studies are obtained by means of Ritz' variational principle [19].

When studying the observable properties of fermion systems represented by many-body states as Eq. (1), one has to evaluate the expectation values of various operators. Although the single-particle states in the FMD wave function are not orthonormal, the expectation values can easily be evaluated by means of a matrix formalism as

$$\mathscr{B}_I = \sum_{pq=1}^{A} \langle q_p|\hat{\mathcal{B}}_I|q_q\rangle \mathbf{o}_{qp}, \qquad \mathscr{B}_{II} = \frac{1}{2}\sum_{pqrs=1}^{A} \langle q_p q_r|\hat{\mathcal{B}}_{II}|q_q q_s\rangle (\mathbf{o}_{qp}\mathbf{o}_{sr} - \mathbf{o}_{qr}\mathbf{o}_{sp}). \tag{2}$$

The wave function's determinant structure is embodied in the matrix $\mathbf{o}$, representing the inverse of the overlap matrix $\mathbf{n}$ with $\mathbf{n}_{pq} = \langle q_p|q_q\rangle$. These two matrices are fundamental for a correct fermion description. Therefore, and because their eigenvalues can cover many orders of magnitudes, it is paramount to calculate these matrices as accurately as possible (i.e. analytically).

FMD was originally devised to describe heave-ion reactions in a time-dependent framework [15, 16]. Now, FMD is also successfully used to describe stationary situations related to nuclear structure and nuclear reactions by mixing many Slater determinants [19, 20]. However, due to the long-range character of the Pauli correlations, evaluating bulk fermion matter becomes a tedious task. The dimension of the matrices usually impedes simulations of a large number of particles, however desirable these may be for bulk fermion systems such as crustal matter. Applying FMD on bulk fermion systems has therefore been a long open question. This has recently been solved within the concept of "periodic boundary conditions" by making the spatial positioning of the single-particle states periodic [21, 22]. An infinite fermion system is created by replicating a unit cell—containing A particles—on an infinite lattice $\mathfrak{B}$. Each cell can be identified by a lattice vector $\boldsymbol{R} = n_1\boldsymbol{a}_1 + n_2\boldsymbol{a}_2 + n_3\boldsymbol{a}_3$ with integer n_j. Using the translation operator $\hat{\mathcal{T}}(\boldsymbol{R})$, the FMD many-body state can be written as

$$|Q_\infty(\boldsymbol{z}_1,\ldots,\boldsymbol{z}_A)\rangle = \hat{\mathcal{A}} \bigotimes_{\boldsymbol{R}\in\mathfrak{B}} \hat{\mathcal{T}}(\boldsymbol{R}) \left\{|q_1(\boldsymbol{z}_1)\rangle \otimes \cdots \otimes |q_A(\boldsymbol{z}_A)\rangle\right\}. \tag{3}$$

As indicated earlier, the overlap matrix $\mathbf{n}$ and its inverse $\mathbf{o}$ are fundamental for the correct fermion description of the system. Although both matrices have infinite dimensions, they exhibit a peculiar nested block-Toeplitz structure which can be exploited [21]. As a result of the translation invariance of the system, each $A \times A$ block of the overlap matrix can be identified unambiguously with the lattice vector $\boldsymbol{R}$ that connects the two cells of the bra and ket states. The blocks can be evaluated as $\mathbf{n}_{pq,\boldsymbol{R}} = \langle q_p|\hat{\mathcal{T}}(-\boldsymbol{R})|q_q\rangle$. Although $\mathbf{n}$ and $\mathbf{o}$ are of infinite dimension, it is possible to obtain the inverse overlap matrix through the following scheme

$$\boldsymbol{\mathcal{N}}(\boldsymbol{k}) = \sum_{\boldsymbol{R}\in\mathfrak{B}} \mathbf{n}_{\boldsymbol{R}}\, e^{-i\boldsymbol{k}\cdot\boldsymbol{R}}, \qquad \boldsymbol{\mathcal{O}}(\boldsymbol{k}) = \boldsymbol{\mathcal{N}}(\boldsymbol{k})^{-1}, \qquad \mathbf{o}_{\boldsymbol{R}} = \frac{1}{V_{BZ}}\int_{BZ} \boldsymbol{\mathcal{O}}(\boldsymbol{k})\, e^{i\boldsymbol{k}\cdot\boldsymbol{R}}\, \mathrm{d}^3\boldsymbol{k}, \tag{4}$$

where BZ represents the first Brillouin zone of the lattice $\mathfrak{B}$ and $\boldsymbol{k}$ is a vector in this volume [21].

The strength of the proposed formalism reveals itself upon evaluating the operators in reciprocal space. Normally, the expectation values of the infinite fermion system would be calculated by means of Eqs. (2) resulting in infinite sums over the block structure. These sums, however, translate into integrals over the first Brillouin zone which are computationally straightforward to evaluate. The expectation value of a one-body operator per unit-cell volume can then be computed as

$$\mathscr{B}_{\rho,I} = \frac{1}{V_{BZ}}\int_{BZ} \sum_{pq=1}^{A} \boldsymbol{\mathcal{B}}_{I,pq}(\boldsymbol{k})\boldsymbol{\mathcal{O}}_{qp}(\boldsymbol{k})\, \mathrm{d}^3\boldsymbol{k}, \qquad \boldsymbol{\mathcal{B}}_{I,pq}(\boldsymbol{k}) = \sum_{\boldsymbol{R}\in\mathfrak{B}} \langle q_p|\hat{\mathcal{B}}_I\hat{\mathcal{T}}(\boldsymbol{R})|q_q\rangle\, e^{i\boldsymbol{k}\cdot\boldsymbol{R}}. \tag{5}$$

Journal of Physics: Conference Series **342** (2012) 012011 doi:10.1088/1742-6596/342/1/012011

Here we assume that the operator $\hat{\mathcal{B}}_I$ commutes with the translation operator $\hat{\mathcal{T}}(\boldsymbol{R})$ over any lattice vector $\boldsymbol{R}$. The two-body operator has a similar structure.

3. Results

When studying properties of bulk matter, it is important that the constraints of the simulation do not interfere with the result and that the properties of the simulation should be equivalent with those of the macroscopic system. It stands to reason that for non-periodic systems, the periodicity should not influence the simulation's results. As long as the correlation volume of the interaction does not exceed the simulation volume, the imposed periodicity works fine. However, serious problems arise in the presence of long-range correlations such as induced by the Pauli exclusion principle. We demonstrate that the proposed technique reproduces two essential features intrinsic to free fermions. The single-particle states are Gaussian wave packets of the form $\langle \boldsymbol{x}|a\boldsymbol{b}\rangle = \exp\{-(\boldsymbol{x}-\boldsymbol{b})^2/(2a)\}$ where the complex vector $\boldsymbol{b}$ represents the mean phase-space position and a is a complex parameter connected with the spreading of the wave packet in phase space.

In Ref. [15] it was shown that, for a hundred periodically positioned wave packets, Eqs. (2) reproduce the momentum and spatial densities intrinsic to a one-dimensional Fermi system. An identical result was achieved in Ref. [21] by using periodic boundary conditions with a unit cell containing a single particle. The densities were evaluated using Eqs. (5). In this reference, it was shown that with increasing overlap of neighbouring wave packets, the effect of antisymmetrisation makes the fermionic behaviour apparent. This effect is also seen in Fig. 1 which compares the kinetic energy of the same periodic system with that of an individual Gaussian and that of a free one-dimensional Fermi gas as function of the wave packet spreading. Although this one-dimensional case hints towards free fermion behaviour, it must be mentioned that, in one dimension, the free Fermi gas results and that of a weak periodic system coincide when having identical particle densities. In more dimensions, it can be shown that the one-particle-in-a-box results evolve towards those of a weak periodic system [21]. This effect is expected because with increasing overlap, ergo increasing wave packet spreading, the wave packet seems uniform to the

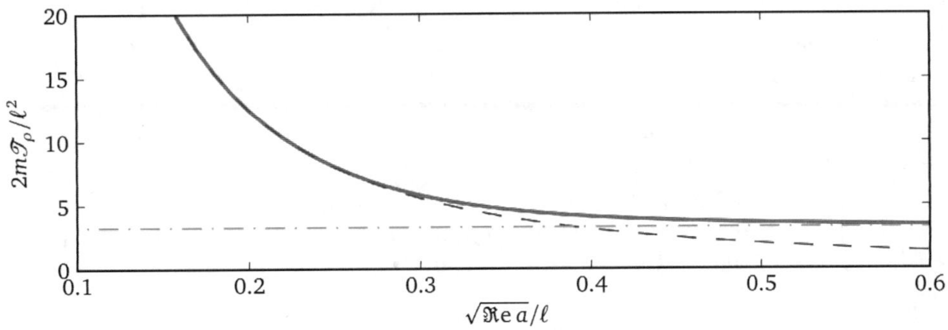

Figure 1. The kinetic energy of periodic Gaussian wave packets (solid line) is compared with that of a single Gaussian (dashed line) and the free Fermi gas (dash-doted line). The kinetic energy per particle $\mathcal{T}_\rho$ is scaled with the mass m of the fermion and the periodic spacing ℓ (box-size) of the Gaussians. The kinetic energy is presented as a function of the scaled variance $\sqrt{\Re\mathrm{e}\, a}/\ell$ of the Gaussian wave packets.

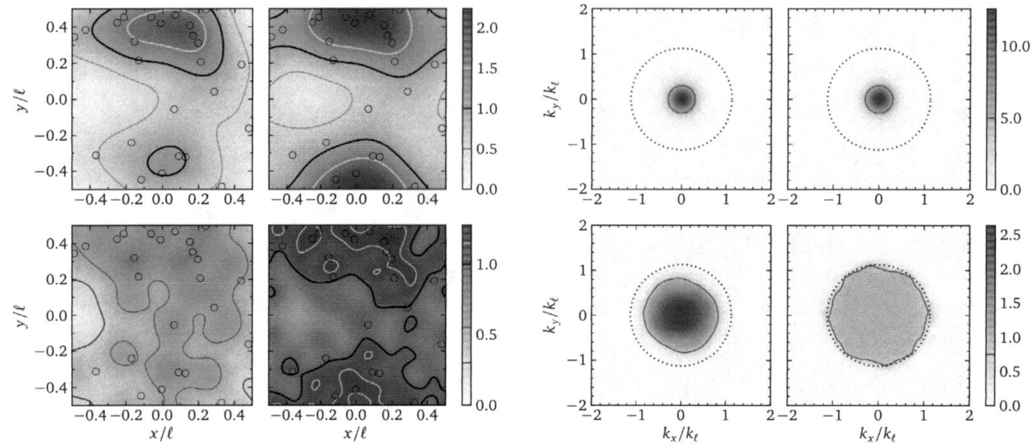

Figure 2. Comparison of the scaled density distributions (left four panels: spatial $\rho_x \ell^2/N$; right four panels: momentum $\rho_k k_\ell^2/N$) of $N = 25$ Gaussian wave packets under different conditions. The wave packets are randomly placed in the unit cell of a square lattice with unit length ℓ and have a width parameter given by $a = (0.2\ell)^2$. The mean momentum of the individual wave packets is set to zero. The spatial densities are normalised and coordinates are expressed in units of ℓ. Analogously for the momentum density with coordinates expressed in units of $k_\ell = \sqrt{N}\pi/\ell$. For each type of density distribution: the upper-two panels depict distinguishable particles, the lower-two panels represent the antisymmetrised case, the left and right panels show the effect of a calculation without and with periodicity, respectively, as described in the text. For the spatial density, the black circles show the positions of the centroids of the wave packets. For the momentum density, the dotted circle represents the Fermi "sphere" of a system of free particles with the same mean density, $k_F = \sqrt{4\pi N}/\ell$. Note the difference in colour scaling.

unit cell and the system behaves as if it consisted of plane waves in a weak periodic system. A successful investigation of bulk matter obviously requires that influences of the geometry of the chosen boundary conditions are negligible, and thus the wave packets should be smaller then the unit cell.

Fig. 2 depicts the effect of periodicity and antisymmetry on the spatial (left hand side) and momentum distributions (right hand side) for 25 single-particle states with zero mean momentum randomly distributed in a square unit cell. For distinguishable particles (upper four panels) the momentum distribution of the 25 particles is the same as the one given by the individual wave packet. When enabling the periodic boundary conditions, only the spatial density grows at the boarders due to the tails leaking in from the neighbouring cells. For indistinguishable fermions (lower four panels), where the many-body wave function is antisymmetrised, one sees that the spatial density (without periodic boundary conditions) in the lower left most panel is much smoother. The fermions tend to avoid closeness in coordinate space and at the same time enlarge the occupied momentum space as indicated in the corresponding panel for the momentum density, thus, inducing Fermi motion. When enabling the periodic boundary conditions, the spatial density increases again due to the tails of the neighbouring cells. In momentum space, the distribution is even more spread due to the antisymmetrisation which is now also with respect to the particles in all surrounding cells, neighbouring and beyond. One obtains an almost uniform Fermi distribution where deformations in the surface reflect the clustering of the spatial density. Hence, a simulation of clustered densities with the proposed periodic structure reproduces intrinsic bulk fermion behaviour, unaffected by the imposed periodic structure which leaves no trace in the momentum distribution. This clearly shows that antisymmetry, a long-range many-body correlation, should not be ignored or truncated for the bulk description.

4. Conclusion

This proceedings gave a short overview how periodic boundary conditions can be implemented to study fully antisymmetrised, infinitely extended, inhomogeneous fermion matter by means of localised single-particle states. The bulk matter is created by spatially distributing a finite number of wave packets in a unit cell which is periodically repeated over space. By treating each single-particle state as an individual state, the overlap matrix and its inverse are of infinite size but possess a nested Toeplitz structure. This is a feature that can be exploited to reduce the computational cost and eliminate the infinite dimensionality of the equations. This allows us to study bulk fermion matter by means of a finite number of fermions. The structure of the resulting equations to evaluate expectation values, resembles those of a finite fermion system, but requires an extra integration over the first Brillouin zone. The later is a translation of the periodic boundary conditions. Although the equations only address a finite number of particles in a unit cell, they keep track of the Fermi statistics of the infinite system. It is demonstrated that this formalism incorporates intrinsic bulk fermion behaviour and reproduces density distributions and energies of free fermion systems. Furthermore, the use of localised states makes the technique suitable to study inhomogeneous fermion matter, such as the neutron star's crust, by means of molecular dynamics or Monte Carlo methods. The next step is to include the forces among the fermions to study the expected foam-like pasta structures.

Acknowledgments

This work was supported by the Helmholtz International Center for FAIR (HIC for FAIR) within the framework of the LOEWE program launched by the State of Hesse (Germany).

References

[1] Chamel N and Haensel P 2008 *Living Rev. Relativity* **11** 10
[2] Rüster S B, Hempel M and Schaffner-Bielich J 2006 *Phys. Rev. C* **73** 035804
[3] Haensel P, Potekhin. A Y and Yakovlev D G 2007 *Neutron Stars 1 : Equation of State and Structure* (Astrophysics and Space Science Library vol 326) (Berlin : Springer)
[4] Ravenhall D G, Pethick C J and Wilson J R 1983 *Phys. Rev. Lett.* **50** 2066–2069
[5] Sonoda H, Watanabe G, Sato K, Yasuoka K and Ebisuzaki T 2008 *Phys. Rev. C* **77** 035806
[6] Oyamatsu K, Hashimoto M and Yamada M 1984 *Prog. Theor. Phys.* **72** 373–375
[7] Horowitz C J, Pérez-García M A, Carriere J, Berry D K and Piekarewicz J 2004 *Phys. Rev. C* **70** 065806
[8] Watanabe G, Sonoda H, Takiwak T, Sato K, Yasuoka K and Ebisuzaki T 2009 *Phys. Rev. Lett.* **103** 121101
[9] Nakazato K, Oyamatsu K and Yamada S, 2009 *Phys. Rev. Lett.* **103** 132501
[10] Newton W and Stone J R 2009 *Phys. Rev. C* **79** 055801
[11] Avancini S S, Barros Jr. C C, Menezes D P and Providência C 2010 *Phys. Rev. C* **82** 025808
[12] Allen M P and Tildesley D J 1991 *Computer simulations of liquids* (Oxford : Clarendon)
[13] Aichelin J and Stöcker H 1986 *Phys. Lett. B* **176** 14–19
[14] Maruyama T, Niita J and Iwamoto A 1996 *Phys. Rev. C* **53** 297–304
[15] Feldmeier H and Schnack J 2000 *Rev. Mod. Phys.* **72** 655-688
[16] Ono A and Horiuchi H 2004 *Prog. Part. Nucl. Phys.* **53** 501–581
[17] Horowitz C J and Kadau K 2009 *Phys. Rev. C* **102** 191102
[18] Dorso C, Duarte S and Randrup J 1987 *Phys. Lett. B* **188** 287–294
[19] Neff T and Feldmeier H 2008 *Eur. Phys. J. Special Topics* **156** 69–92
[20] Neff T 2011 *Phys. Rev. Lett.* **106** 042502
[21] Vantournhout K, Jachowicz N and Ryckebusch J 2011 *Phys. Rev. C* **84** 032801
[22] Vantournhout K, Neff T, Feldmeier H, Jachowicz N and Ryckebusch J 2011 *Prog. Part. Nucl. Phys.* **66** 271–275

Journal of Physics: Conference Series **342** (2012) 012012 doi:10.1088/1742-6596/342/1/012012

Symmetry energy within the BHF approach

Isaac Vidaña[1†], Constança Providência[1§], A. Polls[3‡] and Arnau Rios[3¶]

[1]Centro de Física Computacional. Deparment of Physics, University of Coimbra, PT-3004-516 Coimbra (Portugal)

[2]Departament d'Estructura i Constituents de la Matèria and Institut de Ciències del Cosmos, Universitat de Barcelona, Avda. Diagonal 647, E-08028 Barcelona, Spain

[3]Department of Physics, Faculty of Engineering and Physical Sciences, University of Surrey, Guildford, Surrey GU2 7XH, United Kingdom

E-mail: †ividana@fis.uc.pt

E-mail: §cp@teor.fis.uc.pt

E-mail: ‡artur@ecm.ub.es

E-mail: ¶A.Rios@surrey.ac.uk

Abstract. We analyze the correlations of the slope and curvature parameters of the symmetry energy with the neutron skin thickness of neutron-rich isotopes, and the crust-core transition density in neutron stars. Microscopic Brueckner–Hartree–Fock results are compared with those obtained with several Skyrme and relativistic mean field models. Our results confirm that there is an inverse correlation between the neutron skin thickness and the transition density.

1. Introduction

Isospin asymmetric nuclear matter is present in nuclei, especially far away from the stability line, and in astrophysical systems, particularly in neutron stars. Therefore, a well-grounded understanding of the properties of isospin-rich nuclear matter is a necessary ingredient for the advancement of both nuclear physics and astrophysics. However, some of these properties are not well constrained yet. In particular, the density dependence of the symmetry energy $E_{sym}(\rho)$ is still an important source of uncertainties. Its value J at saturation is more or less well established ($\sim$ 30 MeV), and its behavior below saturation is now much better known [1]. However, for densities above ρ_0, $E_{sym}(\rho)$ is not well determined yet, and the predictions from different approaches strongly diverge. Why $E_{sym}(\rho)$ is so uncertain is still an open question whose answer is related to our limited knowledge of the nuclear force, and in particular of its spin and isospin dependence [2, 3, 4, 5, 6, 7, 8, 9]. Fortunatelly, a major effort is being carried out to study experimentally these properties, and experiments at CSR (China), FAIR (Germany), RIKEN (Japan), SPIRAL2/GANIL (France) and FRIB (USA), can probe the behavior of the symmetry energy close and above saturation density [10]. Additional information on $E_{sym}(\rho)$ can be extracted from the astrophysical observations of compact objects which open a window into both the bulk and the microscopic properties of nuclear matter at extreme isospin asymmetries [11]. In this work, we analyze the correlations of the slope, $L = 3\rho_0 \partial E_{sym}/\partial\rho|_{\rho_0}$, and curvature, $K_{sym} = 9\rho_0^2 \partial^2 E_{sym}/\partial\rho^2|_{\rho_0}$, parameters of the symmetry energy with the neutron skin thickness of neutron-rich isotopes, and the crust-core transition density in neutron stars. The results are obtained within the microscopic Brueckner–Hartree–Fock (BHF) approach, and are compared with those obtained with several Skyrme and relativistic mean field (RMF) models.

2. The BHF approach of asymmetric nuclear matter

Assuming charge symmetry of nuclear forces, the energy per particle of asymmetric nuclear matter can be well approximated in terms of the isospin asymmetry parameter, $\beta = (N - Z)/(N + Z) = (\rho_n - \rho_p)/\rho$, as

$$\frac{E}{A}(\rho, \beta) \sim E_{SNM}(\rho) + E_{sym}(\rho)\beta^2 \ , \tag{1}$$

where $E_{SNM}(\rho)$ is the energy per particle of symmetric nuclear matter, and $E_{sym}(\rho) = E/A(\rho, \beta = 1) - E_{SNM}(\rho)$ is the so-called symmetry energy.

It is common to characterize the density dependence of the energy per particle of symmetric matter around the saturation density ρ_0 in terms of a few bulk parameters by expanding it in a Taylor series around ρ_0,

$$E_{SNM}(\rho) = E_0 + \frac{K_0}{2}\left(\frac{\rho - \rho_0}{3\rho_0}\right)^2 + \frac{Q_0}{6}\left(\frac{\rho - \rho_0}{3\rho_0}\right)^3 + \mathcal{O}(4) \ . \tag{2}$$

The coefficients denote, respectively, the energy per particle, the incompressibility coefficient and the third derivative of symmetric matter at saturation,

$$E_0 = E_{SNM}(\rho = \rho_0) \ , \quad K_0 = 9\rho_0^2 \frac{\partial^2 E_{SNM}(\rho)}{\partial \rho^2}\Big|_{\rho=\rho_0} \ , \quad Q_0 = 27\rho_0^3 \frac{\partial^3 E_{SNM}(\rho)}{\partial \rho^3}\Big|_{\rho=\rho_0} \ . \tag{3}$$

Similarly, the behaviour of the symmetry energy around saturation can be also characterized in terms of a few bulk parameters,

$$E_{sym}(\rho) = J + L\left(\frac{\rho - \rho_0}{3\rho_0}\right) + \frac{K_{sym}}{2}\left(\frac{\rho - \rho_0}{3\rho_0}\right)^2 + \frac{Q_{sym}}{6}\left(\frac{\rho - \rho_0}{3\rho_0}\right)^3 + \mathcal{O}(4) \ , \tag{4}$$

where J is the value of the symmetry energy at saturation and the quantities L, K_{sym} and Q_{sym} are related to its slope, curvature and third derivative, respectively, at such density,

$$L = 3\rho_0 \frac{\partial E_{sym}(\rho)}{\partial \rho}\Big|_{\rho=\rho_0} \ , \quad K_{sym} = 9\rho_0^2 \frac{\partial^2 E_{sym}(\rho)}{\partial \rho^2}\Big|_{\rho=\rho_0} \ , \quad Q_{sym} = 27\rho_0^3 \frac{\partial^3 E_{sym}(\rho)}{\partial \rho^3}\Big|_{\rho=\rho_0} \ . \tag{5}$$

The BHF approach of asymmetric nuclear matter starts with the construction of all the G matrices describing the effective interaction between two nucleons in the presence of a surrounding medium. They are obtained by solving the Bethe–Goldstone equation

$$G_{\tau_1\tau_2;\tau_3\tau_4}(\omega) = V_{\tau_1\tau_2;\tau_3\tau_4} + \sum_{ij} V_{\tau_1\tau_2;\tau_i\tau_j} \frac{Q_{\tau_i\tau_j}}{\omega - \epsilon_i - \epsilon_j + i\eta} G_{\tau_i\tau_j;\tau_3\tau_4}(\omega) \tag{6}$$

where $\tau = n, p$ indicates the isospin projection of the two nucleons in the initial, intermediate and final states, V denotes the bare NN interaction, $Q_{\tau_i\tau_j}$ the Pauli operator that allows only intermediate states compatible with the Pauli principle, and ω, the so-called starting energy, which corresponds to the sum of non-relativistic single-particle energies of the interacting nucleons. The single-particle energy ϵ_τ of a nucleon with momentum $\vec{k}$ is given by

$$\epsilon_\tau(\vec{k}) = \frac{\hbar^2 k^2}{2m_\tau} + Re[U_\tau(\vec{k})] \ , \tag{7}$$

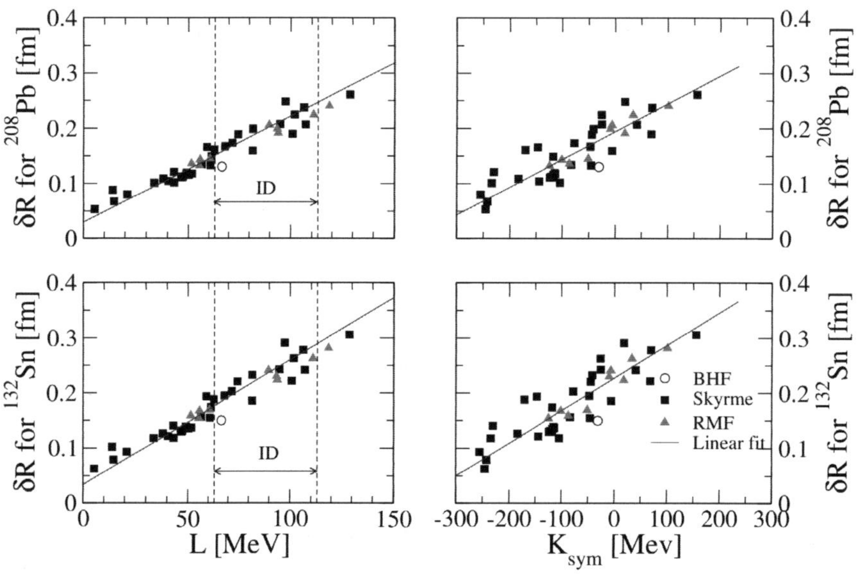

Figure 1. Neutron skin thickness for ^{208}Pb (upper panels) and ^{132}Sn (lower panels) versus L (left panels) and K_{sym} (right panels). The vertical dashed lines on the left panels denote the constraints on L from isospin diffusion (ID) [10].

where the single-particle potential $U_\tau(\vec{k})$ represents the mean field "felt" by a nucleon due to its interaction with the other nucleons of the medium. In the BHF approximation, $U(\vec{k})$ is calculated through the on-shell energy G-matrix, and is given by

$$U_\tau(\vec{k}) = \sum_{\tau'} \sum_{|\vec{k}'|<k_{F_{\tau'}}} \langle \vec{k}\vec{k}' \mid G_{\tau\tau';\tau\tau'}(\omega = \epsilon_\tau(k) + \epsilon_{\tau'}(k')) \mid \vec{k}\vec{k}' \rangle_A \quad (8)$$

where the sum runs over all neutron and proton occupied states, and the matrix elements are properly antisymmetrized. Once a self-consistent solution of Eqs. (6) and (8) is achieved, the energy per particle of asymmetric matter (and consequently the symmetry energy) can be calculated as

$$\frac{E}{A}(\rho, \beta) = \frac{1}{A} \sum_{\tau} \sum_{|\vec{k}|<k_{F_\tau}} \left(\frac{\hbar^2 k^2}{2m_\tau} + \frac{1}{2} Re[U_\tau(\vec{k})] \right) . \quad (9)$$

We note here that the BHF calculation carried out in this work uses the realistic Argonne 18 [12] two-body potential and the Urbana IX [13] three-body force. For the use in the BHF calculation the three-body-force has been reduced to a two-body density dependent force by averaging over the coordinates of the third nucleon [14]. For further reading, and details on the Skyrme forces and the relativistic models considered in this work, the reader is referred to Ref. [15].

3. Results

It has been shown by Brown and Typel [16], and confirmed later by other authors, that the neutron skin thickness, $\delta R = \sqrt{\langle r_n^2 \rangle} - \sqrt{\langle r_p^2 \rangle}$, calculated in mean field models with either non-relativistic or relativistic effective interactions is very sensitive to the density dependence of the nuclear symmetry energy, and, in particular, to the slope parameter L at the normal nuclear

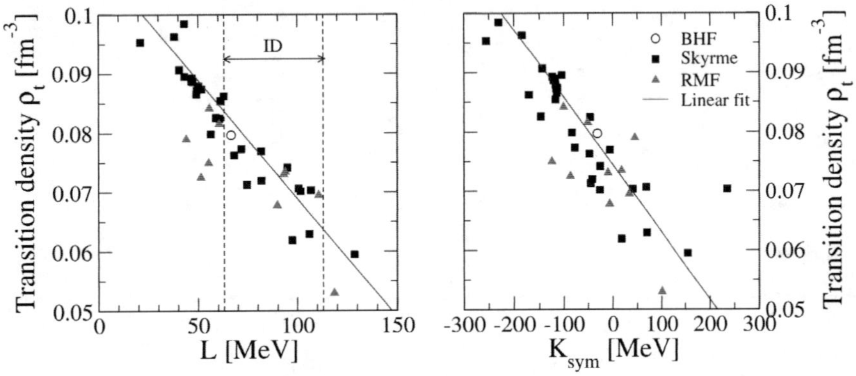

Figure 2. Transition density from non-uniform to uniform β-stable matter as a function of L (left panel) and K_{sym} (right panel). The vertical dashed lines on the left panel denote the constraints on L from isospin diffusion (ID) [10].

saturation density. Using the Brueckner approach and the several Skyrme forces and relativistic models considered, we have made an estimation of the neutron skin thickness of ^{208}Pb and ^{132}Sn and we have studied its correlation with the slope L and curvature K_{sym} parameters. In Fig. 1 we show the correlation between δR for ^{208}Pb and ^{132}Sn with the parameters L (left panels) and K_{sym} (right panels). It can be seen, as has already been shown by other authors, that both the Skyrme forces and the relativistic models predict values of δR that exhibit a tight linear correlation with L. Note that the microscopic Brueckner calculation is in excellent agreement with this correlation. The linear increase of δR with L is not surprising since the thickness of the neutron skin in heavy nuclei is determined by the pressure difference between neutrons and protons, which is proportional to the parameter L, $P(\rho_0, \beta) \sim L\rho_0\beta^2/3$ [17].

Another sensitive quantity to the symmetry energy is the transition density ρ_t from non-uniform to uniform β-stable matter which may be estimated from the crossing of the β-equilibrium equation of state with the thermodynamical spinodal instability line. As it has been shown in Ref. [18] the predictions for the transition density from the thermodynamical spinodal are $\sim 15\%$ larger than the value obtained from a Thomas–Fermi calculation of the pasta phase. Therefore, we expect that our estimation of the transition density from the thermodynamical spinodal will define an upper bound to the true transition density [19]. We display in Fig. 2 ρ_t as a function of the parameters L and K_{sym} for the BHF calculation together with the predictions of the several Skyrme forces and relativistic models. It is clear from the figure that ρ_t is sensitive to the slope and curvature parameters L and K_{sym} of the symmetry energy, decreasing almost linearly with increasing L and K_{sym} in agreement with recent results [20, 21]. Using the experimental constraint on L from isospin diffusion, we estimate the transition density to be between 0.063 fm^{-3} and 0.083 fm^{-3}. This range is in reasonable agreement with the value of $\rho_t \approx 0.08$ fm^{-3} often used in the literature.

Finally, we show in Fig. 3 the transition density ρ_t from non-uniform to β-stable matter as a function of the neutron skin thickness in ^{208}Pb (left panel) and ^{132}Sn (right panel) for our Brueckner calculation and the different Skyrme forces and relativistic models. The figure shows, as already pointed out by Horowitz and Piekarewicz [22], that there is an inverse correlation between the neutron skin thickness and ρ_t. In [22] a Walecka model with non-linear $\omega - \rho$ terms was used and the transition density was obtained with an RPA approach. We confirm the same trend for a larger set of nuclear models. Note that, again, our microscopic Brueckner results are in very good agreement with this correlation. As pointed out in Ref. [22], these

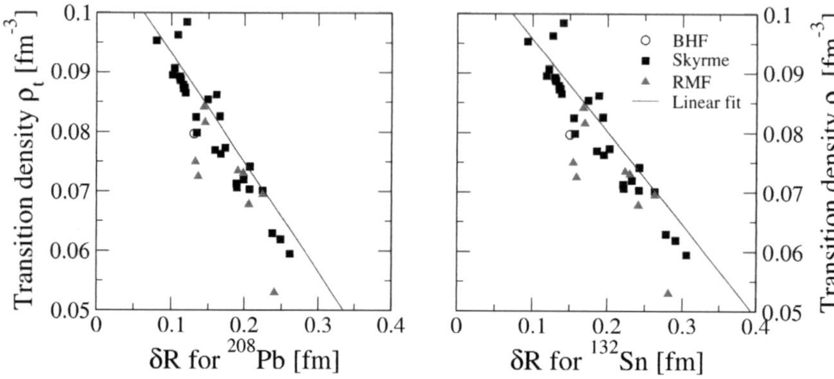

Figure 3. Transition density from non-uniform to uniform β-stable matter vs the neutron skin thickness for ^{208}Pb (left panel) and ^{132}Sn (right panel).

results suggest that an accurate measurement of the neutron radius in heavy nuclei like ^{208}Pb or ^{132}Sn is absolutely essential since it can provide considerable and valuable information on the properties of neutron star's crust.

4. Summary

Summarizing, we have studied the correlation between the neutron skin thickness of neutron-rich isotopes and the parameters L and K_{sym}. We have found that the BHF results are in very good agreement with the correlations already predicted by other authors using non-relativistic and relativistic effective models. We have also analyzed the correlations of L and K_{sym} with the transition density ρ_t from the crust to the core in cold neutron stars. Using the experimental constraint on L from isospin diffusion, we have estimated the value of ρ_t to be between 0.063 fm^{-3} and 0.083 fm^{-3}, a range in reasonable agreement with the the value of $\rho_t \approx 0.08$ fm^{-3} often used in the literature. Finally, we have confirmed for a large set of nuclear models that there is an inverse correlation between the neutron skin thickness and the transition density ρ_t, a trend pointed out first by Horowitz and Piekarewicz in Ref. [22].

Acknowledgments

This work has been partially supported by FEDER and FCT (Portugal) under the projects PTDC/FIS/113292/2009, CERN/FP/109316/2009 and CERN/FP/116366/2010, the Consolider Ingenio 2010 Programme CPAN CSD2007-00042 and Grant No. FIS2008-01661 from MEC and FEDER (Spain) and Grant 2009GR-1289 from Generalitat de Catalunya, a Marie Curie Intra European Fellowship within the 7th Framework programe, STFC grant ST/F012012, and by COMPSTAR an ESF Research Networking Programme.

References

[1] M. B. Tsang *et al.* Phys. Rev. Lett. 102, 122701 (2009); M. B. Tsang *et al.*, Prog. Part. Nucl. Phys. **66**, 400 (2011).
[2] V. R. Pandharipande, V. K. Garde, Phys. Lett. B **39**, 608 (1972).
[3] R. B. Wiringa, V. Fiks and A. Fabrocini, Phys. Rev. C **38**, 1010 (1988).
[4] G. E. Brown and M. Rho, Phys. Lett. B **237**, (1990).
[5] I. Bombaci and U. Lombardo, Phys. Rev. C **44**, 1892 (1991); W. Zuo, I. Bombaci and U. Lombardo, Phys. Rev. C **60**, 024605 (1999); W. Zuo, A. Lejeune, U. Lombardo and J. F. Mathiot, Eur. Phys. J. A **14**, 469 (2002).
[6] A. E. L. Dieperink, Y. Dewulf, D. Van Neck, M. Waroquier and V. Rodin, Phys. Rev. C **68**, 064307 (2003).

[7] C. Xu and B. A. Li, Phys. Rev. C **81**. 064612 (2010).
[8] C. Xu and B. A. Li, arXiv:1104.2075v1 (2011).
[9] F. Sammarruca, Phys. Rev. C 84, 044307 (2011).
[10] L. W. Chen, C. M. Ko and B. A. Li, Phys. Rev. Lett. **94**, 032701 (2005); Phys. Rev. C **72**, 064309 (2005); B. A. Li, L. W. Chen and C. M. Ko, Phys. Rep. **464**, 113 (2008).
[11] A. W. Steiner, M. Prakash, J. Lattimer and P. J. Ellis, Phys. Rep. **411**, 325 (2005).
[12] R. B. Wiringa, V. G. J. Stoks, and R. Schiavilla, Phys. Rev. C **51**, 38 (1995).
[13] B. S. Pudliner, V. R. Pandharipande, J. Carlson and R. B. Wiringa, Phys. Rev. Lett. **74**, 4396 (1995).
[14] B. A. Loiseau, Y. Nogami, and C. K. Ross, Nucl. Phys. A **165**, 601 (1971); **176**, 665(E) (1971); P. Grangé, M. Martzolff, Y. Nogami, D. W. L. Sprung, and C. K. Ross, Phys. Lett. B **60**,237 (1976); M. Baldo and L. Ferreira, Phys. Rev. C **59**, 682 (1999).
[15] I. Vidaña, C. Providência, A. Polls and A. Rios, Phys. Rev. C **80**, 045806 (2009).
[16] B. A. Brown, Phys. Rev. Lett. **85**, 5296 (2000); S. Typel and B. A. Brown, Phys. Rev. C **64**, 027302 (2001).
[17] P. Danielewicz and J. Lee, Nucl. Phys. A **818**, 36 (2009).
[18] S.S. Avancini, L. P. Brito, J.R. Marinelli, D.P. Menezes, M.M.W. Moraes, C. Providência and A. M. Santos, Phys. Rev. C **79**, 035804 (2009).
[19] C. J. Pethick, D. G. Ravenhall and C. P. Lorenz, Nucl. Phys. A **584**, 675 (1995); F. Douchin and P. Haensel, Phys. Lett. B **485**, 107 (2000).
[20] J. Xu, L.W. Chen, B.A. Li and H.R. Ma, Astrophys. J. 697, 1549 (2009)
[21] K. Oyamatsu and K. Iida, Phys. Rev. C **75**, 015801 (2007).
[22] C. J. Horowitcz and J. Piekarewicz, Phys. Rev. Lett. **86**, 5647 (2001).

Journal of Physics: Conference Series **342** (2012) 012013 doi:10.1088/1742-6596/342/1/012013

The influence of magnetic field geometry on magnetars X-ray spectra

D Viganò[1], N Parkins[2,3], S Zane[3], R Turolla[4,3], J A Pons[1] and J A Miralles[1]

[1] Departament de Fisica Aplicada, Universitat d'Alacant, Ctra. San Vicent del Raspeig, s/n. 03690 San Vicent del Raspeig, Spain

[2] University of Liverpool, Liverpool L69 3BX, UK

[3] Mullard Space Science Laboratory, University College London, Holmbury St. Mary, Dorking, Surrey RH5 6NT, UK

[4] Università di Padova, Dipartimento di Fisica, via F. Marzolo 8, I-35131 Padova, Italy

E-mail: daniele.vigano@ua.es

Abstract. Nowadays, the analysis of the X-ray spectra of magnetically powered neutron stars or *magnetars* is one of the most valuable tools to gain insight into the physical processes occurring in their interiors and magnetospheres. In particular, the magnetospheric plasma leaves a strong imprint on the observed X-ray spectrum by means of Compton up-scattering of the thermal radiation coming from the star surface. Motivated by the increased quality of the observational data, much theoretical work has been devoted to develop Monte Carlo (MC) codes that incorporate the effects of resonant Compton scattering (RCS) in the modeling of radiative transfer of photons through the magnetosphere. The two key ingredients in this simulations are the kinetic plasma properties and the magnetic field (MF) configuration. The MF geometry is expected to be complex, but up to now only mathematically simple solutions (self-similar solutions) have been employed. In this work, we discuss the effects of new, more realistic, MF geometries on synthetic spectra. We use new force-free solutions [14] in a previously developed MC code [9] to assess the influence of MF geometry on the emerging spectra. Our main result is that the shape of the final spectrum is mostly sensitive to uncertain parameters of the magnetospheric plasma, but the MF geometry plays an important role on the angle-dependence of the spectra.

1. Introduction

Neutron stars (NSs) can be powered by different sources of energy: rotation, accretion, residual heat and magnetic field. Up to now, more than twenty X-ray pulsating sources are identified as isolated, slowly rotating ($P \sim 2 - 10$ s) neutron stars, whose large X-ray luminosity $L_x \sim 10^{33} - 10^{35}$ erg/s cannot usually be explained in terms of rotational energy, at variance with rotation-powered pulsars (RPPs). Historically, they are classified as Anomalous X-ray Pulsars (AXPs) or Soft Gamma Repeaters (SGRs), and they are recognized as magnetars. In these objects, MF decay and evolution is the dominant energy source and crustal deformations produced by the huge internal magnetic stresses are the likely cause of bursts and episodic giant flares. In most cases, the spin-down inferred external magnetic field turns out to be super-strong. Recently, increasing evidence gathered in favor of a unified scenario for isolated neutron stars,

Journal of Physics: Conference Series **342** (2012) 012013 doi:10.1088/1742-6596/342/1/012013

their different observed manifestations being mainly due to a different topology and strength of the internal MF, and to evolution ([11]).

The typical soft X-ray (< 10 keV) spectrum of magnetars is given by a thermal component ($kT \sim 0.3 - 0.5$ keV), plus a hard tail phenomenologically described by a power law with a photon index $\Gamma \sim 3 - 4$ [7]. This hard tail is believed to be produced by resonant compton (up)scattering (RCS) of photons emitted by the star surface onto mildly relativistic particles flowing in magnetosphere. In order to take into account for this effect, several codes have been developed in recent years [4, 8, 9, 5]. The two basic inputs are the external MF and a prescription for the particle space and velocity distribution. With these ingredients, and with a given energy distribution for the incoming surface photons (the seed spectrum), the MC simulations of RCS provide the emerging, reprocessed spectrum, which must be compared with observational data.

Magnetars also exhibit a high-energy ($\sim$ 20–200 keV) power-law tail, the origin of which is still under debate. One of the possible mechanisms is again RCS, whose efficiency at high energy is very sensitive to MF and plasma velocity. In this paper we discuss the effect of the geometry of the MF, which is obtained from new calculations of magnetostatic, force-free equilibria.

2. Force-free twisted magnetosphere

The shape of the internal MF of NSs is largely unknown, but it is thought to be complex, with toroidal and poloidal components of similar strength, and with significant deviations from a dipolar geometry. In the crust, a large toroidal field is a necessary ingredient to explain the observed thermal and bursting properties of magnetars, even in the case in which they show a low value of the inferred external dipolar field [13]. The complex internal topology must be smoothly matched to an external solution, so that the internal MF geometry at the star surface has a strong effect on the external MF. In this framework, the RPP phenomenolgy is expected to be better explained by the presence of a NS with a simple, nearly potential, dipolar field, while magnetars activity is compatible with the presence of a complex external field, as a result of the transfer of magnetic helicity from the internal to the external field. A non-potential MF needs supporting currents, and charges can be either extracted from the surface or produced by one-photon pair creation in the strong MF [1].

In the inner magnetospheric region MF lines corotate with the star, with footprints connected with the crustal field, which evolves on long timescales ($10^3 - 10^5$ years). Furthermore, resistive processes in the magnetosphere act on a typical timescale of years [2], much longer than the typical response of the tenuous plasma, whose Alfvén velocity is near to the velocity of light. For these reasons, a reasonable approximation is to build stationary, force-free magnetospheric solutions, assuming that the footprints are anchored in the crust and neglecting resistivity. Then, the ideal MHD force-free equation is a simple balance between electromagnetic forces, provided that inertial, pressure, and gravity terms are negligible

$$\rho_e \vec{E} + \frac{1}{c}\vec{J} \times \vec{B} = 0 \,, \tag{1}$$

where we have introduced the rotationally induced electric field $\vec{E} = -\vec{v}_{rot} \times \vec{B}$, the related charge density ρ_e and the current density $\vec{J} = c\vec{\nabla} \times \vec{B}/4\pi$.

The electric field gives corrections of order $(\rho\Omega/c)^2$. Neglecting such contributions is safe if we consider the region closest to a magnetar surface ($r \lesssim 100 r_\star$), and remembering that the light cylinder for slow rotators lies typically at a few thousands stellar radii ($r_l = c/\Omega$ and magnetar periods are in the 1–10 s range). Therefore, condition (1) reduces simply to $\vec{J} \parallel \vec{B}$: the particles can flow only along MF lines. Even if the pulsar equation simplifies in this limit, it still contains an arbitrary function, which basically describes the toroidal field as a function of the poloidal one (see [14] for the mathematical formulation). In the literature, only the simplest (semi-)analytical solutions have been derived and later employed in applications. In

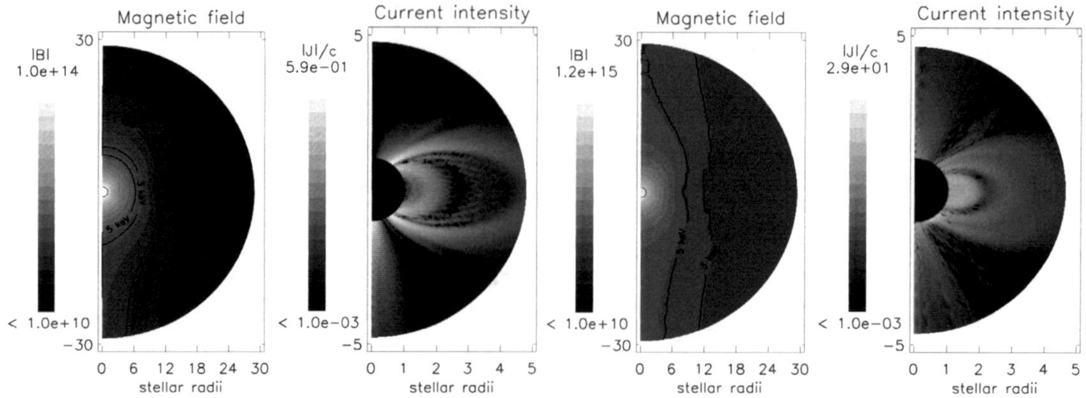

Figure 1. Numerical model 1 (left) and model 2 (right). First panel: $|\vec{B}|$ (grey logarithmic scale) with scattering surfaces for photons of 1, 3 and 5 keV (here $\beta = 0.5$, $\mu = 0.5$; see text); second panel: current intensity $|\vec{J}|/c$ (grey linear scale), in units of 10^{14} G$/r_\star$.

particular, in the context of magnetars, the self-similar twisted dipole family of solutions is the most popular choice. Proposed for solar corona studies [6], and later extended to the magnetar scenario [12], in self-similar models the radial dependence of the MF components is described by a simple power law, $B_i \propto r^{-(p+2)}$. These solutions are semi-analytical and described by only one parameter, the radial index p (see [10] for globally twisted multipolar fields). However, globally twisted solutions necessary evolve in time and magnetic configurations in which the twist affects only a limited bundle of field lines are likely required by observations [2].

As an alternative to self-similar models, in [14] we presented a numerical method to build general solutions. The force-free solutions are obtained by iteration starting from an arbitrary, non-force-free poloidal plus toroidal field. By the introduction of a fictitious velocity field proportional to the Lorentz force, the code is designed to dissipate only the part of the current not parallel to the MF. The boundary conditions are an arbitrary, fixed radial MF at the surface, $B_r(r_\star, \theta)$, and the requirement of a continuous matching with a potential dipole at large distances $r_{out} \geq 100 r_\star$. This method is able to produce general configurations, whose features may be radically different from the self-similar solutions. In this work, we try to reproduce some j-bundle like structure, aiming at providing input for MC simulations [9]. In particular, we focus on two models, whose MF and current distribution are shown in Fig. 1. The first one is obtained fixing a dipolar form for radial MF, while the second one has a concentration of field lines near the south pole. To get these profiles, we have chosen the ϕ component of the vector potential as:

$$\text{model 1:} \qquad A_\phi(\theta, r_\star) = B_0 \frac{\sin\theta}{2} \tag{2}$$

$$\text{model 2:} \qquad A_\phi(\theta, r_\star) = B_0 \frac{\sin\theta}{2} \left[1 + 4e^{-\frac{(\theta - 0.9\pi)}{0.2^2}} \right] \tag{3}$$

from which we recover $B_r = (\vec{\nabla} \times A_\phi \hat{\phi})_r$ (B_0 is the value of B_r at north pole). Both models have high helicity, but it is not homogeneously distributed like in self-similar models. Model 1 is characterized by an equatorial asymmetry, with current and twist concentrated in a closed bundle near the equatorial region and a j−bundle near the southern semi-axis. Model 2 is more extreme: its helicity is $\sim$ 25 times larger than for model 1. Both are supposed to be more realistic than the self-similar models, since currents are more concentrated near the axis.

Journal of Physics: Conference Series **342** (2012) 012013 doi:10.1088/1742-6596/342/1/012013

3. Resonant Compton scattering

The MC code we employ is presented in [8, 9] and we refer to these works for further details. Here we just summarize some basic concepts. It is assumed that currents flow in a non-potential magnetosphere and that the electron density is large enough for the medium to be optically thick to resonant scattering: photons produced by the cooling star surface can be then up-scattered by RCS, populating the hard tail of the spectrum. In the stellar frame, the resonant energy corresponds to the Doppler-shifted gyration frequency of the electron

$$\omega_D(B,\beta) = \frac{1}{\gamma(1-\beta\mu)}\frac{eB}{m_e c} \tag{4}$$

where $B = |\vec{B}|$ is the local intensity of MF, $c\beta$ the electron velocity, $\gamma = 1/\sqrt{1-\beta^2}$ the Lorentz factor, and μ the cosine of the angle between the photon direction and the electron momentum. The non-resonant contributions are negligible, since the resonant cross-section is up to five orders of magnitude larger than the Thomson cross-section. Figs. 1 show the scattering surfaces in the two models, for photons of 1, 3 and 5 keV, assuming $\beta = 0.5$ and $\mu = 0.5$. The surfaces are far from being spherically symmetric. In model 2, which has the strongest helicity, the scattering surfaces of soft X-ray photons lie tens of stellar radii away from the surface, especially near the axis.

The scattering optical depth for a given seed photon energy depends on the local particle density n_e, the velocity of the scatterer and the local value of the MF. Since photons can be pushed to energies of some hundreds keVs, non-conservative scattering and the fully relativistic QED cross sections are used [9]. The velocity distribution is assumed to be a one-dimensional (along the field line), relativistic Maxwellian distribution with a plasma temperature T_e and a bulk velocity β_{bulk}, as described in [8]. The strong simplification in this approach is that the velocity distribution does not depend on position, which reduces the microphysical inputs to two parameters: the temperature of the plasma T_e and β_{bulk}. Only recently, a more consistent treatment of the charges space and velocity distribution was attempted [3]. These new results show that the velocity distribution is non-trivial and strongly depends on position, which opens a new range of possibilities, so far unexplored.

Given a seed spectrum for the photons emitted from the surface (assumed as a blackbody with temperature $T_\star$), the MC code follows the photon propagation through the scattering magnetosphere. When a photon enters in a parameter region where no more resonant scatterings are possible, the photon escapes to infinity, where its energy and direction are stored. The sky at infinity is divided into patches, so that viewing effects can be accounted for (e.g. if line-of-sight is along the angles θ_s, ϕ_s, only photons collected in the patch which contains that pair of angles are analyzed). The angle-averaged spectrum is obtained simply by averaging over all patches.

4. Results and discussion

In previous works [15], self-similar magnetospheric models were used to obtain synthetic spectra that were satisfactorily fitted to X-ray observations, giving typical values of $\beta_{bulk} \in [0.1, 0.7]$ and $\Delta\phi \in [0.4, 2.0]$. In this paper we explore the dependence of the spectrum on the magnetosphere model by comparing results obtained with self-similar models and with our new numerical solutions for force-free magnetospheres. We present results comparing models with a magnetic field intensity at the pole of $B_p = 10^{14}$ G. The seed spectrum is assumed to be a $kT_\star = 0.4$ keV Planckian isotropic distribution for both ordinary and extraordinary photons. In Figs. 2 and 3 we plot the angle-averaged spectra, i.e., the distribution of all photons escaped to infinity, independently on the final direction. Fig. 2 shows the effects produced by changes in the bulk velocity, keeping fixed the MF configuration (model 1), and the electron temperature ($kT_e = 20$ keV). The main fact is the the spectrum in the region $E > 10$ keV becomes harder as

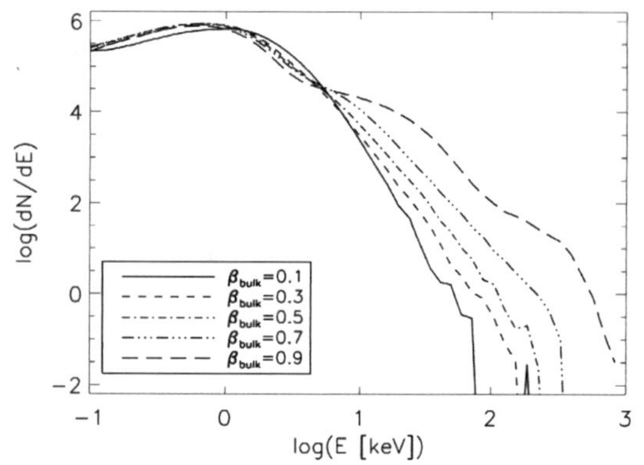

Figure 2. Angle-averaged spectra for $kT_e = 20$ keV, model 1, changing β_{bulk}.

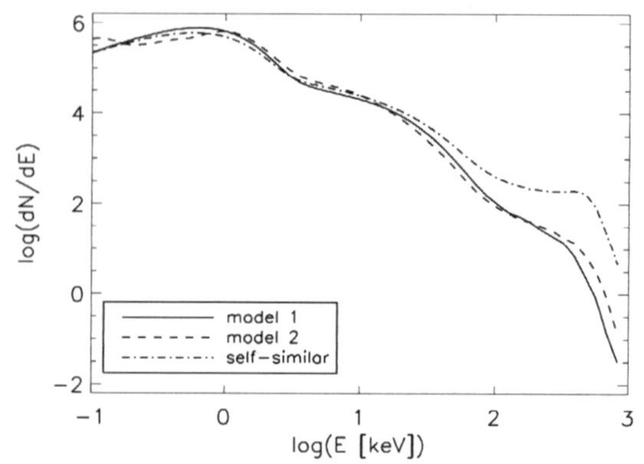

Figure 3. Angle-averaged spectra for $kT_e = 20$ keV, $\beta_{bulk} = 0.9$, changing MF model.

β_{bulk} is increased, while it is depleted of photons of energies in the $1 - 10$ keV range. A similar, but less pronounced effect, is obtained by increasing the electron temperature [9].

In Fig. 3 we compare spectra obtained with fixed values of $kT_e = 20$ keV and $\beta_{bulk} = 0.9$, varying only the MF topology. The three lines correspond to model 1 (solid), model 2 (dashes) and a self-similar model (dash-dotted line) with $\Delta\phi = 1.36$, which approximately has the same total helicity of model 1. We have chosen a high value of β_{bulk} to show a case where the effects are larger, but our conclusions do not change qualitatively for lower values of β_{bulk}. The high-energy cutoff produced by electron recoil is also evident. We see that the major differences arising from changes in MF topology appear in the hard tail of the spectra, which is clearly harder for the self-similar model. However, the thermal part of the spectrum is more depleted, especially for model 2.

However, in the angle-averaged spectrum, the differences due to the MF topology are partially smeared out, and the comparison between synthetic spectra as seen from different viewing angles (first three panles in Fig. 4) is more interesting. The comptonization degree reflects the inhomogeneous particle density distribution and therefore the particular current distribution produces important differences between different viewing angles. Since neutron stars rotate, the study of pulse profiles and phase-resolved spectra can trace the geometric features of the scattering region (tens of stellar radii). In the lower right panel we show the light curves for the three MF models, in bands 0.5–10 keV and 20–200 keV, for an oblique rotator with a certain line of sight ($\xi = \chi = 30^\circ$ in [8]).

For self-similar models (lower left panel), the thermal part of the spectrum shows a smooth dependence with the viewing angle, which in turn translates into relatively regular light curves. In models 1 and 2, differences in the spectra induced by viewing angles are more pronounced. In these cases, the spectrum is much more irregular and asymmetries are larger. In particular, model 1 (upper left) shows a softer spectrum when seen from northern latitudes, with important spectral differences. The pulsed fraction of model 1 is very high in the hard X-ray band, while it is comparable with the self-similar model in the soft range. Model 2 (upper right) has a more symmetric distribution of currents, and the comptonization degree at different angles depends on the energy band in a non trivial way. This results in large pulsed fraction for both energy ranges and in notable differences between their pulse profiles.

The high energy tail, seen at different colatitudes, can vary by one order of magnitude or more. Investigating the geometry can help to recognize the different components seen in the hard tails via pulse phase spectroscopy. On the other hand, the macrophysical approach used in this work to obtain the MF topology should be accompanied by the corresponding microphysical

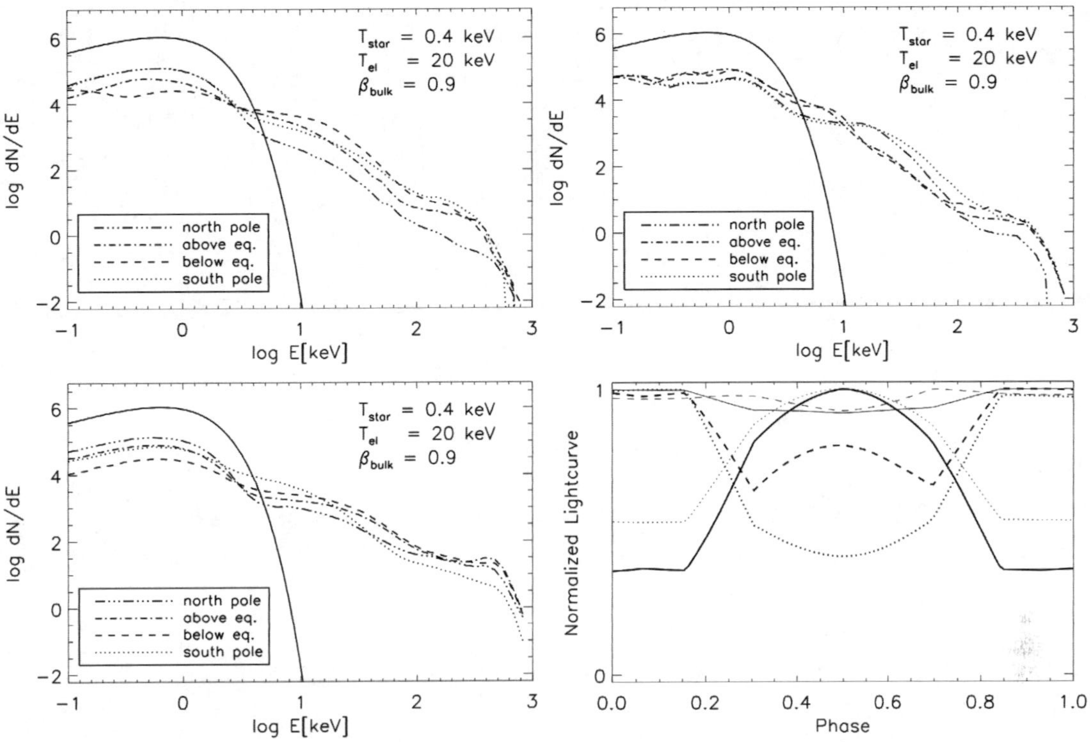

Figure 4. Synthetic spectra for model 1 (upper left panel), model 2 (upper right), self-similar model (lower left) with the parameters indicated in the figures. Legends indicate the four viewing angles here considered: aligned with north or south pole, just above and just below the equator. The seed blackbody is shown for comparison (solid line). Fourth panel shows the pulse profiles obtained for model 1 (solid line), model 2 (dotted), self-similar (dashed), in the energy ranges 0.5–10 keV (thin lines) and 20–200 keV (thick lines), with $\xi = \chi = 30°$ (see text).

description to determine the velocity distribution of particles, here simply taken as a model parameter. A fully consistent description of both MF geometry and the spatial and velocity distribution of particles is needed to advance our interpretation of magnetar spectra.

References

[1] Beloborodov A M and Thompson C 2007 *ApJ* **657** 967
[2] Beloborodov A M 2009 *ApJ* **703** 1044
[3] Beloborodov A M 2011 *High-Energy Emission from Pulsars and their Systems* ed Torres D F and Rea N p 299
[4] Fernández R and Thompson C 2007 *ApJ* **660** 615
[5] Fernández R and Davis S W 2011 *ApJ* **730** 131
[6] Low B C and Lou Y Q 1990 *ApJ* **352** 343
[7] Mereghetti S 2008 *A&ARv* **15** 225
[8] Nobili L, Turolla R and Zane S 2008 *MNRAS* **386** 1527
[9] Nobili L, Turolla R and Zane S 2008 *MNRAS* **389** 989
[10] Pavan L, Turolla R, Zane S and Nobili L 2009 *MNRAS* **395** 753
[11] Pons J A and Perna R 2011 *ApJ* **741** 123
[12] Thompson C, Lyutikov M and Kulkarni S R 2002 *ApJ* **574** 332
[13] Turolla R, Zane S, Pons J A, Esposito P and Rea N 2011 *ApJ* **740** 105
[14] Viganò D, Pons J and Miralles J A 2011 *A&A* **533** A125
[15] Zane S, Rea N, Turolla R and Nobili L 2009 *MNRAS* **398** 1403

AUTHOR INDEX